Resilience:

Stories of Grit, Grace, and Ascent to the C-Suite

Resilience:

Stories of Grit, Grace, and Ascent to the C-Suite

itsmf INFORMATION TECHNOLOGY SENIOR MANAGEMENT FORUM

INFORMATION TECHNOLOGY SENIOR MANAGEMENT FORUM
PUBLISHING

itsmf INFORMATION TECHNOLOGY
SENIOR MANAGEMENT FORUM

Published by:

Information Technology Senior Management Forum

Itsmfleaders.org

@itsmfleaders

First Edition: June 2024

Book Designer: Dawn Jordan Jones | 29Eleven Media

ISBN 978-1-7365068-7-5 (paperback)
ISBN 978-1-7365068-8-2 (ebook)

Table of Contents

Foreword

by Dr. James O. Rodgers, Director, Executive Academy

It has been my good fortune, over my career, to work directly with and to observe dozens of senior executive leaders (Chairmen, CEOs, Presidents), when I was an employee, a consultant, a large-scale change lead, a trusted advisor, a friend, and an executive coach. As an engineer and a self-described observer, I make a point to notice and document patterns of experience and behavior that precede a person's elevation to the highest levels of organizational leadership. Over the years, I have codified those observations and have committed myself to sharing that code with young executives who have an interest in executive leadership.

In summary, that code says, "It is not what you know (technical and otherwise). It is not who you know. It is who knows you!" Understanding and practicing this simple idea can change your focus and your experience as a rising executive.

The idea is so simple it's tempting to print it on the back of a card and present it as a single lecture. My experience with adult learning tells me that is a doomed approach. I've seen so many incredibly successful, talented, and renowned executives discover how difficult it is to make real, sustainable change. It can be so easy to attend a six-week lecture series designed to transmit information about leadership and corporate management. My objective, though, is to encourage behavior change so that my students experience different outcomes in their careers.

We all know that an extremely small amount of transmitted information will be adopted as knowledge.

And that adopted knowledge will translate to an even smaller set of learnings (new ways of thinking and being). As Malcolm Knowles, co-author of *The Adult Learner: The Definitive Classic in Adult Education and Human Resource Development* has said, "Learning (alone) produces behavior change." Learning takes time.

The authors of this book have participated in a "slow drip" facilitated learning adventure over ten months. During that time, they have refreshed their knowledge, learned from each other, and practiced new behaviors. The results have been remarkable. One of the principles we subscribe to in Executive Academy is, "Learn, then do." It is based on the ancient spiritual principle "Be ye doers of the word and not hearers only." So, we consume information; we adopt knowledge; and we embrace learning as a prelude to sustainable behavior change.

Leaders tell stories. You will see how the lessons of storytelling are evident in the essays presented in this book. These executive leaders recognize that storytelling is the most enduring form of communicating with the humans one wants to lead. Telling their own personal stories reminds them how unique, memorable, and interesting they are as leaders.

I am thrilled with the content of this book. I encourage anyone who wants to "crack the code" of executive leadership to read and reread these generous essays. You will be blessed.

Part I

Early Years: Building on Solid Foundations

"The fool speaks; the wise man listens."

– Proverbs 12:15

Early Lessons about First Impressions Fuel Long-Term Career Growth

by Jaclyn Parran

I have been exposed to leadership since birth. I am the product of two God-fearing, strong Black parents who poured everything into me. They believed in leading by example and often leveraged teachable moments providing lessons to me and my younger brother. Some of my earliest memories include gathering for family dinners and discussing our days. My parents often spoke about the importance of making a good first impression when meeting others. The rationale for this, they explained, is because you never know who God may be sending your way to pass along a blessing. I certainly never wanted to miss any of my blessings, but as an introvert, meeting new people was an activity I often dreaded. Anytime I was placed in a situation that required an introduction to someone new, my parents would prepare me by reminding me to "give a firm handshake" and "speak up so your voice can be heard."

My parents understood that there are times where you may only receive one chance to make a good first impression. That meant it was important to *always* leave a good first impression. Little did I know that this advice would become permanently ingrained, and often called upon when establishing relationships. I would eventually learn that positive first impressions are crucial when networking in professional environments. They are also important when establishing yourself and building your brand. In the early years of building my career, first impressions served as the catalyst to collaborate with team members, identify mentors, and secure sponsors. Today, I share this same advice with my children and those that I lead and mentor.

Making a good first impression.
Working at a weekend fruit stand in an outdoor central Florida flea market at the age of 15 taught me the importance of embracing every opportunity. Receiving positive feedback for pleasant customer service through customer tips and additional cash rewards from the owner of the fruit stand, reinforced the learnings from my parents and stressed the significance of making a good first impression.

Seven years later, as a bright-eyed, eager to learn, recent college graduate, I approached my first role in corporate America with the same energy as that 15-year-old. I moved 1,000 miles from my home in sunny central Florida to the Midwest to experience a new life full of opportunities—and changes in climate four times a year. I embraced my new normal through lighthearted conversations and laughter when meeting many colleagues over the course of several years. Often, I would even be told, "You made a great first impression on (a mutual colleague of ours), so I am looking forward to getting to know you." Hearing those words then and hearing them now makes me smile.

At this stage of my career, I have a full appreciation for how first impressions can help grow or hinder your career and I've developed an acronym—FUN—that sums up a consistent approach to making a good first impression.

- *F - Find the Connection.* When meeting someone for the first time, identify what you may have in common to break the ice and jumpstart the discussion. This might be family, pets, hobbies, travel.
- *U - Unique.* As you chat, incorporate information that is thought-provoking and will make the discussion memorable. Look for opportunities to highlight any assistance you may be able to provide to that person or someone within their network, based on your knowledge and/or skills.
- *N - Next Steps.* Based on your initial conversation, are you interested in speaking further? If so, see if you can determine the individual's preferred method of follow-up communication (email, text, phone call, internal company network). Before the discussion concludes, come to an agreement about a future meeting and take responsibility for

sending a calendar invitation even before you part ways. Scheduling while still in the presence of the individual demonstrates that you have enjoyed the conversation and signals your willingness to prioritize getting to know that person better.

Making good first impressions can lead to long-lasting friendships, solid mentorships, and confirmation of sponsorships. As my parents always said, "Be kind always, and if you can help people, you should." Living by the motto has paid dividends both personally and professionally.

First impressions can lead to mentorship.

As my career has grown, I have benefited from the wisdom of internal and external mentors. Many of those mentoring relationships (as the mentor or the mentee), were born through positive first impressions. Internal mentors have helped me understand my brand, prepare for internal promotions, and navigate company politics. External mentors have assisted with broadening my perspective, providing industry insights, and challenging me to accelerate and continue to think differently.

One of my first internal corporate mentoring relationships formed organically after I met my mentor through a program team that required us to work on the same deliverables. I approached our introductory meeting in a "FUN" manner and my mentor matched my energy and seemed equally authentic, agreeing to provide help and guidance whenever I needed it. As a seasoned veteran of the company and a senior member of the program team, my mentor began to educate me on how processes were developed and how decisions were truly made within the team and the organization. He also provided positive reinforcement on the tasks I completed and introduced me to other colleagues beyond the program team and our organization. From him I gained many valuable insights which allowed me to establish a strong foundation within the company.

While benefiting from the relationship, I was also doing my best to keep him informed of all of the data points I was learning that were relevant to *his* success—insights from water cooler conversations and lunch meetings with others. My willingness to share valuable information with my mentor helped reassure him that I was not only focused on my own growth and development, but also invested in his. He and I have not worked together for several years now, but he still makes himself available to assist me with solutioning challenges when needed and I do the same for him. Although we do not share the same demographic background or life experiences, he has taught me through words and actions what great mentorship looks like. From this experience, I learned to approach all mentoring relationships with a level of intentionality and purpose. Just as food, water, and pruning are required for plants and flowers to blossom, mentoring relationships require care and feeding to continue to properly grow into something mutually beneficial.

As I continued my career, I encountered additional mentors who have been outstanding and designated many of these esteemed and diverse leaders as my *personal board of directors*, which remains busy and active even when I do not always realize it. Each member of my board has been strategically identified based on an encounter that has allowed us to have a transparent and beneficial interaction. They have positively pushed me, helped me to think critically, taught me how to advocate for myself in the corporate world, assisted me with overcoming challenges with direct reports, peers, and senior leaders, and also helped me with parenting, prepared me for new opportunities, and more.

In 2017, several members stepped in to help me weigh the pros and cons of a large career move. As a result of a major workforce restructuring initiative, I was offered the role as leader of a cyber technical team. Prior to this, my career had consisted of leadership roles within information technology that closely aligned with business integrations and operations. I was intrigued by the new role because it presented an opportunity to further develop leadership skills by driving delivery and outcomes within a cyber technical team.

At the time, several thoughts were top of mind: 1) I was not a subject matter expert in any of the processes or tools for which the cyber team was responsible; 2) While I had tenure with the company, I would be required to work in an organization with leaders and direct reports who I did not know well and who did not know of me because they had been with the company for two years or less; 3) My husband's role required him to travel 80% of each month; 4) Our son was 18 months old; 5) I was pregnant with our daughter; 6) During the week, my husband and I had limited child care support beyond daycare; and 7) I was several months into the IT Senior Management Forum's Inaugural EMERGE Academy my company was sponsoring me to attend.

Conversations with my personal board involved prayer, whiteboarding, question and answer sessions, coffee shop discussions (with my toddler son in tow), and at times even house calls. Although all were not convinced that moving into the opportunity would be best for my career (and sanity!), I prayed about it and ultimately decided to accept. Once I communicated my decision to my personal board, they shifted to advisors, intent on helping me succeed and demonstrated support every step of the way. The journey of leading and learning in that role while navigating constant change in my personal life was not easy; however, I was reminded of my parents' lesson: "With God, all things are possible." Not only did that role allow me to further develop personally and professionally, but it provided many experiences that have encouraged me to be the confident wife, mother, daughter, leader, sponsor, and mentor I am today.

Start with making a good first impression and you will be surprised at the relationships that flourish. True mentorship only happens when you connect with people on a deeper level. To do that you have to go beyond the surface. Surface connections may open doors, but real relationships will give you the discernment and guidance needed for excellence. Have some FUN!

Jaclyn Parran is a Business Information Security Officer, Manufacturing and Quality at Eli Lilly and Company. A strategic visionary leader with nearly 20 years of information technology and cybersecurity experience, she is a powerful force known for inspiring teams to attain operational excellence and helping leaders achieve their potential.

Unveiling the Magic Within: Embracing Authenticity, Personal History, and Storytelling

by Alonna Davis

There are moments in life, in our careers, where we lose our bearings. Surrounded by all the things we prayed for, we may look around at "success" and feel like a stranger to ourselves. In moments of silent reflection, you may wonder, "How did I get here?" and "Is this where I want to be?"

I've been in that position. Even though I have acquired a lot of titles and accolades over the years, with each new beginning, with each new phase in my career, I have had to find my magic all over again. There's still a lot I don't know—about life, my career, what my future might hold—but as I stand adrift, feeling lost, I do know one thing with deeply rooted certainty: I know who I am. Amidst life's twists and turns, I have been able to plant my feet on that rock and find myself again, more times than I can count. There's magic in finding yourself and knowing who you are; there's power in remembering where you came from.

Remember the hands that shaped you.

My family lived in the Thunderbolt neighborhood of Savannah, Georgia, in a small green house peeling and sagging with age. The house belonged to my grandmother, who lived on the first floor, while my mom, my older brother and I stayed on the second floor. My brother and I grew up sharing a room just large enough to fit two twin-sized beds. Nothing in the green house was new, the floorboards creaked with each step, and the house trembled under the weight of a strong breeze. Everything I treasured had been passed down and once belonged to someone else. By the same token, every sorrow and every triumph were a shared experience; my grandmother's hope and perseverance, a hard-won hand-me-down to my mother and then to me.

My grandmother had a no-nonsense attitude and a penchant for doling out harsh truths with a side of tough love. She wore sensible shoes and kept a pair of reading glasses on a chain around her neck. In the evenings after dinner, she taught us piano, using a ruler to rap our knuckles, admonishing, "Remember your hands!" Looking down, I would quickly correct the positioning of my fingers on the ivory keys, ever aware of her watchful eye—and the ruler! My mom (having lived through her own round of piano lessons) would tap out the cadence of the songs with her foot, grounding us in a consistent rhythm of support. Night after night, I watched my hands marching over scales and gliding over keys, wondering if my fingers would ever grow as long as my mom's, as proficient as my grandmother's.

Over the years I have watched my mother's hands change into my grandmother's hands. They grow veiny and thin, the backs wrinkling as the skin folds atop bent, calloused fingers. These hands that taught me to play "Twinkle Twinkle Little Star," are the same hands that consoled and wiped away tears. These are hands that worked three jobs and bled to keep food on the table. Squeezing my mother's hands now, I think of my grandmother's hands, knitted with strength, and crossed gently over her chest as we closed her casket after saying our final goodbye. I think of the long line of hands holding me up. My childhood was the sound of scales echoing through the house; it was warm hands juxtaposed against a cold unforgiving world. When I feel lost, personally, or professionally, I remind myself of those hands.

Remember the climb.

My older brother and I are not quite two years apart in age. He's got my grandmother's tenacity and my mother's heart—pinned to his sleeve, visible for the world to see. I spent most of my childhood following him around (much to his dismay); there was nowhere he could go that I wouldn't follow. Standing in the doorway of the green house, I would watch as he rode his bike down the worn dirt driveway, so impressed by how he navigated the potholes and divots, subtly shifting to avoid rough terrain and rocks, always good at maneuvering through life. I would like to think I got the latter skill from him.

There's an early family photograph of us sitting on a playground, black and white and worn with age, folded at the edges and well-loved. In the photo, his arms are wrapped protectively around my waist to keep me from sliding off the bench. Once, looking down at the photo, my mom shook her head in loving frustration before saying, "You have always been *his* baby."

She often tells the story of how she left us alone for a minute, only to find me teetering halfway up the stairs in the green house, as I climbed to reach my brother. Focused solely on him, I followed the sound of his voice, one wobbly foot at a time, as he repeated "Come on baby! Come on baby!" Oblivious to the dangers below, I continued to climb with no thought or fear of falling. Three steps left. Two. Then, the upstairs landing finally in sight. Mom recalls how she forgot to breathe and didn't make a sound as she waited for him to grab me in his arms at the top step, celebrating my journey and pulling me away from the edge. She tells the story to remind me of how fearless I have always been, but what resonates is how I have always been supported. Surrounded by love and empowered to dream big without fear of falling. In moments of self-doubt, I hear my brother's voice urging me to take that first step and remind myself that before I could walk, I could, with help, climb.

Remember your magic.

I have had a zigzag career marked by ups and downs. My path has been anything but linear as I continue to reinvent myself. I have been a makeup artist and a director of sales. I've been a lawyer and a senior vice president of global information security. I've been the head of governance, risk and compliance at some of the world's largest financial institutions. Perhaps more significantly, I've been the "only" in more rooms than I can count. While I have achieved, by certain measures, a high level of success and influence, I still sometimes feel insecure and unseen, a work in progress, and I hope that I will always be. Now, as I navigate through privileged spaces, surrounded by power and reeling to find my footing, I am left to find myself all over again. With each new position, I am pushed to remember who I am and to tap into my magic. For me, magic isn't about a title or an achievement; it's about embracing your story and finding the guts to share it with other people.

My magic was born in the cold pre-dawn hours of the morning, as my mother left the green house to clock in at the first of three jobs. Coming home exhausted, she would drop her purse on the wooden side table by the door before coming into the living room to ask my brother and me, "How was your day at school?" My magic was watered by my brother's constant affirmations and grew with every whispered "come on" that (to this day) continues to propel me forward when I feel like giving up. My magic is rooted in the legacy of my grandmother's tireless hands.

When I speak of who I am as a leader, I speak of my mother and her resilience. I speak of my grandmother and her humility. In interviews, very rarely do I tell stories of how I have climbed the corporate ladder, but I often mention my brother and the impact of his support on my journey. I do sometimes share stories of my professional accomplishments, but I always remember to ground them in the people and the moments that have shaped me as a person.

There is magic in your story, and power in sharing it with others.

Alonna Davis is a Governance and compliance Executive at Truist Financial Corporation, leading first-line risk advisory services for technology integration and performance. Her experience spans technology and legal fields, including business process improvement, legal/regulatory requirements, and global strategy development and implementation. Alonna earned her J.D. degree from Vanderbilt University Law School. As a banking executive and attorney, she has helped organizations improve operational risk management, strengthen internal controls, and enabled profitable growth within risk appetites.

Get in the Zone: How Football, an Injury, and Switching Majors Drove Future Career Moves

by Paul Stephens

In this book, my colleagues highlight key concepts for rounding out your early career years, solid advice on navigating the climb through middle management, important guidance for transitioning to senior management, and finally, bold tactics for breaking into the C-Suite. As for me, I'd like to address the crucial and often stormy passage from one stage to the next in the early and middle part of your career in information technology.

My own nugget of advice is simply this: You cannot achieve anything alone. Or, as the English priest and poet John Donne wrote in his work *Meditation XVII*, in 1624: "No man is an island, entire of itself; every man is a piece of the continent, a part of the main."

I invite you to consider this idea.

I often reflect on how my upbringing, culture, friends, religious belief, football teams I've been a part of, education, and work history have contributed to where I am today. I reflect to find patterns and habits I need to break or to determine those that I want to enhance.

At this moment, I am a director in the Cloud and Digital: Salesforce practice of my organization. I got here, I believe, for a few reasons, and I hope this blueprint helps you in your own career growth.

~~~

I grew up in northeast Ohio and in 9th grade, joined the local high school football team, where I played defensive end. Football is how I learned the importance of a team. Football also gave me a goal that was larger than myself. Every individual had a part to play, from
~~~

groundskeeping, to equipment management, the medical staff, coaches, and players.

I dove headfirst into being the best defensive end I could. I kept my grades above the qualifying line, worked out relentlessly, ran with my position teammates, scrutinized film on my opponents to find their weaknesses, studied and practiced our defensive playbook. I worked on hand-to-hand combat techniques to allow me to get past my opponent and drilled with the team. I ate and drank football, laser-focused enough that I won a scholarship to play at the University of Cincinnati (UC), where I continued everything, at a faster pace with more weight and dedicated study time.

Players blessed enough to start or get a lot of playing time have successfully proven they belong in the position. If you get these opportunities consistently, you begin to feel unstoppable—that is the only way I can describe it, that feeling that you are on top, and you can't be toppled. You are "in the zone," in a state of optimal performance, fully immersed in an activity and performing at your peak with a sense of effortless concentration. I believe being in the zone doesn't only apply to the world of sports. Anyone can find the zone, but how?

Legendary football coach Vince Lombardi once said, "The difference between a successful person and others is not a lack of strength, not a lack of knowledge, but rather a lack in will."

Will works like this. First, decide on a path. I found my first path because I found something where I could contribute my whole self without holding back: a community in football. My grades in high school, while not outstanding, had gotten me into the College of Engineering at UC, but my focus on football meant I couldn't balance that program's demands and I had already transitioned to the Business College. Then, my self-perceived greatness was thwarted when I tore my anterior cruciate ligament (ACL) and needed several subsequent surgeries. I needed to find a new vision.

Before football, I was already interested in computer technology. Rather than a future in football, my new goal was to be a technology consultant, to help companies leverage technology to drive digital transformation.

~~~

I got this computer interest from my father. During early junior high years, he would challenge my siblings and me to write code on new Microsoft DOS machines just as Windows was becoming popular. I remembered that feeling of excitement when I made stick figures move on a screen. At UC, I learned that I while I might not be able to come up with new coding languages or discover processes to enhance computing speed and capability as taught in the Engineering College, I did enjoy learning how businesses leveraged technology to increase supply chain efficiencies, categorize and automate business processes, implement statistical models to represent and predict economic trends to forecast where sales might end at the end of a quarter.

I graduated with a degree in Information Systems and learned to tap my network—my old teammates, doctors, counselors, my mother, uncle, and siblings—to help me build and transition into a new vision. Just as I had once sought opportunities to study football positions with help from position coaches who could grow someone who never played in an organized sport before, I needed teachers and mentors to settle into life without football, a life leveraging computer technology to transform corporations to help them become and/or remain competitive.

Two people stand out in my academic life, who catapulted my understanding through exposure and challenge: First, the amazing woman who counseled me to become a (non-football playing) student in the College of Business. Football players did not get the same internship opportunities, and after my injury she challenged me to find new mentors and build a new team.

As a junior in high school, I was one of the top defensive ends in the Steel Valley conference and received contingency offer letters from several Division 1 schools to play football on a full scholarship. However, by the time I graduated, mediocre grades
~~~

meant those offers were no longer available. I remember my mother stepping in to take me and a couple of teammates around the northern Midwest states for interviews at Division 3 schools. But I had a vision for myself that was bigger than Division 3. In summer after my senior year, I was invited to participate in the Steel Valley Conference All-Star game and ended up with the most tackles and tackles for loss at the defensive end position. A reporter for the *Youngstown Vindicator* interviewed me, realized that I hadn't signed with a Division 1 team—and got me an interview with the UC Bearcats. By the end of the summer, the head coach awarded me a scholarship.

I draw a parallel to this story when recalling my next mentor, a fantastic programmer and professor who taught C++ and Java in the College of Business, classes I was required to take. After my ACL tear, I remembered my passion for coding, and I leaned in. This professor taught me the concepts of object-oriented programming, which I used to understand database modeling.
Eventually I became an unofficial tutor in both of these subjects, finding a new place to be in the zone.

After graduation, I got a job as a programmer at a cellular billing company, running programs that generated cell phone bills and detecting fraudulent activations. Then at a career fair a few years later, I met an HR lead for a small consulting company, where I furthered my programming and database skills; but this did not move the needle towards my goal to be a consultant. A few years later though, I was hired by a small Cincinnati-based consulting company which delivered custom code-based solutions for companies in the area.

There, I made friends with skilled coders, learned their trade, moved into different roles, and, when I moved into sales, eventually was mentored by the company founder and CEO who exposed me to elevated conversations with executives. His hobby was cooking, and he shared his love for food by inviting clients to in-person chef how-to events where he would cook foods very different from what I'd grown up eating. The CEO often called me into his office to review my client list, opportunities, and plan for sales in a given quarter (after the sales leader and I formed these plans). During these one-on-ones, he challenged me to read at least two books a quarter and to build personal relationships with clients. To help me with the latter, he had me invite my clients to his cooking experiences. In this way, during my time there, I built a new network of friends to count on, supporting one another. I'd move beyond coding and into selling an idea to organizations and thus, learned the fundamentals of being a consultant.

Then, the market crash and the Great Recession of 2008 meant the company needed to restructure and I and a few others lost our jobs. Just before this, I was blessed to have met a sales rep from Salesforce.com in 2007, who approached my company with a partnership helping move companies to the cloud, targeting small to medium business consulting. I learned as much as I could about Salesforce, got myself certified, and worked with them to sell their solutions to our clients. Though this was not enough to save my job from the recession, a couple of salespeople started a new company, saw my potential, and asked me to help them continue to sell Salesforce as a contractor working for a start-up consulting company.

The co-founders of this new company gave me my start in consulting and for the next two years I helped the team sell and implement Salesforce.com to companies large and small. I had found the zone again! Once I got to the zone—just as I had once wanted to play football in a larger division to increase the challenge and hone my skills further—I wanted to work for a larger consulting firm. Thus, two years later I moved to a large consulting firm as a senior associate, and finally to another even larger consulting firm.

As a director in this last large firm's Cloud and Digital Salesforce practice, I've been implementing digitally transformative solutions for my clients—my goal way back in college, post-injury. My move from senior associate at one large consulting firm to director at another was grounded in earlier lessons and helped by people who saw something in me, trusted me enough to be there for me in the roundtables, believed in me enough to promote me, and pushed me to continue to grow.

Along the way, I've developed a short list of key takeaways I'd like to leave you with, helpful for anyone who wants to move into senior management.

- Find your passion. Reach deep into yourself to identify experiences that move you to believe in something higher than yourself. Notice the blessings given to you.
- Find your team and know your position. As a defensive end, I knew that I alone could never win the game, but that I played a part in reaching that goal.
- Be open: to building new relationships, finding mentors and sponsors, tackling change. You're constantly being groomed for what's next; find people who will have your back.
- Get in the zone. You don't need to know how your vision will play out, but you do need to have a vision. Build your team, share the wins, and own the lows in your career. As Dr. James Rodgers once said, "Build a strategy or a position that moves you into the direction of the vision." Have patience and allow your vision to mature. When you find yourself getting into a mode where you feel amazing at what you love to do, take a moment to study that, challenge it, become the best at it, and then share it with others.

Paul Stephens is Director of PwC's Cloud & Digital: Salesforce practice. Paul brings more than fifteen years of consulting experience selling and delivering large-scale transformation solutions across a variety of industries including aerospace, communications, financial services, insurance, healthcare, high-tech, professional services, power and utilities, and real estate.

Yin & Yang: The Worst and Best Business Advice I Ever Received

by Andrea Pennycooke

"The only thing to do with good advice is to pass it on. It is never of any use to oneself."

– Oscar Wilde

Throughout your career, you are going to receive a lot of advice, most of it unsolicited. I certainly did, especially as I navigated the uncharted waters of management consulting in the early 2000s, fresh out of college with hope in my eyes, dripping with naiveté. When, after 15 years, I hung up my consulting hat and entered slower-paced industry work, the advice from peers, mentors, and managers continued: *Fake it 'til you make it. Don't take risks; stick to what you know. Work harder not smarter. Be open to getting bumps and bruises in new places.*

With all these pearls of wisdom flying around, I found myself wondering: How do I filter through the noise and separate the good advice from the bad? What personal and professional experiences did these individuals navigate that shaped their perspectives? And most important, W*ill this advice work for me?*

Our journeys are not one size fits all and not all advice is good advice. Listening with a discerning ear helped me identify the good and not-so-good nuggets that then served as guiding lights illuminating my path to personal success—and empty words I purposely ignored for fear that they would lead me down a path of treacherous dead ends.

Now, twenty-plus years of hard learned lessons and advice-sifting later, with experience as my guide and a pure heart, I offer to you the worst and best business advice I've ever received. May these contribute to your personal success and more importantly, may you pass it on!

Worst advice ever: ***Fake it 'til you make it.***
From a very young age, my parents taught me the importance of being comfortable in my own skin, displaying humble confidence, knowing and playing to my strengths, being aware of my weaknesses and doing the work necessary to turn them into growth opportunities. They never advised me to fake anything. So, when it comes to "Fake it 'til you make it," I can safely say I have never done this, preferring to be authentic in both personal and professional settings.

I have witnessed peers and leaders fall prey to the lure of this advice, and the outcome was almost always negative. In a world where confidence is often hailed as the key to success, this mantra has gained significant traction. While the concept may offer a temporary boost, relying solely on this approach can have detrimental effects on personal growth and well-being.

Faking it until you make it promotes the idea of projecting confidence outwardly, even when one may not feel it internally. While confidence is important, pretending to have skills or knowledge you don't possess can lead to mistakes and damage your credibility. This facade can create a superficial sense of self-assurance, masking underlying insecurities and impeding genuine personal development.

Instead, I offer up two alternatives that counter this common tongue in cheek mantra.

Instead, try: Learn it and you'll earn it.
Throughout my career, I've been blessed to be surrounded by many positive examples of what "great" looks like—authentic individuals who put in the work to hone their technical and interpersonal skills. These individuals created a safe space for me to learn and grow into the leader that I am today. If I needed to ask a peer who was viewed

as a subject matter expert to complete a knowledge transition session on a new application or software platform that I wasn't familiar with, or shadow a leader at a level that I aspired to in order to understand the role, or simply request direct feedback, I used these interactions to build my own personal action plan for professional growth. In other words, I learned and kept learning, earning confidence.

So, take stock of your abilities. Instead of portraying an idealized version of yourself, cultivate self-acceptance, acknowledge your strengths, weaknesses, and imperfections. What do you believe you are lacking and how can you increase skills in that specific area? Who within your sphere of influence—mentors, sponsors, peers—are operating at the level you aspire to? What are they doing that you aren't? Armed with the answers to these questions, you can then…

Instead, try: Face it so you make it.

Once you have identified growth areas, put in the work. Don't rely on the illusion of confidence. Adopt a growth mindset and focus on the value of continuous learning and resilience in the face of setbacks. Success is hard but rewarding and often requires relentless dedication to facing difficult situations that in turn push you to your limits mentally and physically. But you're built for this! Face the challenge head on. Every hurdle you overcome and every milestone you achieve will contribute to a sense of accomplishment that is deeply gratifying and will help you develop a humble confidence that you won't ever have to fake.

Best advice ever: ***Be open to getting bumps and bruises in new places.***

I can remember it all so clearly: the height of the COVID-19 pandemic, and I was leading an effort for the roll-out of my company's Paycheck Protection Program (PPP) response, a role that would catapult me to the ranks of leadership within our enterprise technology organization. Hours were long and the environment was high stress and fast-paced, but I was happy to be doing this all from the comfort of my own home. Even better, I was working in a domain familiar to me, doing what I loved the most—program management and delivery. Unbeknownst to others, I was also three months pregnant with my second child. Life was good! One could say that I was (that dreaded word) *comfortable*, coasting my way through the current season of my career.

Then, the opportunity arose for a visible leadership role, reporting directly to a Divisional CIO. I was both surprised and nervous. Enter the self-doubt. Was I ready? This was all new territory to me, a domain I wasn't familiar with, supporting business partners I had never worked with. Was my family ready for the stress this would put on precious family time? As I was pregnant, my husband was protective of my mental and physical health and our time together. Enter the wise words of my soon-to-be-manager, someone who remains one of my biggest supporters and mentor: "Andrea, you could always say no. But I think it's time for you to be open to getting bumps and bruises in new places." Mind blown!

What I learned was that while bumps and bruises may seem like setbacks, they often hold valuable lessons and opportunities for personal growth. By reframing these experiences, you can navigate unfamiliar terrain with confidence and resilience. Entering new spaces can be intimidating, but it's also where growth happens. Embrace the unknown as an opportunity to expand your horizons and discover hidden talents and strengths. That's how it worked out for me.

My first three months were filled with interactions with business and technology professionals who had years of marketing experience and were well versed in that language. While I had a wealth of delivery experience and the ability to build relationships, I had zero domain knowledge. Buzz words and concepts flew over my head. Faking it wasn't an, as I knew that would lead to a lack of credibility when, not if, they realized that I had no earthly idea what I was talking about. So, I gave myself grace and used each interaction as a growth opportunity.

I openly admitted my lack of knowledge but promised myself that I would put in the work over the next few months to come up to speed and ensure I was contributing value to my team, partners, and the company. This promise resulted in me spending the last trimester of my pregnancy, during the pandemic, and while working full-time, attaining an online Certificate in Digital Marketing & Selling from the University of Georgia's Terry College of Business. I knew the certification would increase my domain knowledge and help me be a better, more well-rounded leader in my new role. This investment in myself paid dividends, as my partners and team made many references to my ability to hold my own in conversations around the company's strategic direction.

Mistakes are inevitable when exploring new territories. Instead of dwelling on them, see them as valuable learning experiences. Every stumble is an opportunity to gain insight and refine your approach. Each challenge overcome strengthens your ability to navigate future obstacles with grace and determination. Embrace resilience as a powerful tool for personal development. Most importantly, be kind to yourself as you navigate unfamiliar territory. Acknowledge your efforts and celebrate small victories along the way. Self-compassion fosters resilience and empowers perseverance through difficult times.

Embracing bumps and bruises in new places was not just good advice—it's essential for personal growth and development. Experience showed me that every bump and bruise is a badge of honor, a testament to courage and resilience in the face of adversity. Go and step boldly into new spaces, get some bumps and bruises of your own, each one bringing you one step closer to the best version of yourself, personally and professionally.

Andrea Pennycooke is head of Marketing, Digital Onboarding, and Workflow Automation Technologies at Truist Financial Corporation. A technology executive with over 20 years of experience, Andrea has spent her career building high-performing, cross-functional teams that bridge the gap between technology execution and business outcomes, to drive transformational change.

Beyond the Badge: Cultivating Meaningful Relationships and Navigating Social Terrain with Lessons Learned from My Mom, the Detective

by DaVon Gorman

The art of cultivating meaningful relationships is often overlooked and undervalued, yet stands as the cornerstone of my own success. When I reflect on the most critical attribute of my professional and personal achievements, I see that relationship building, recognizing the contributions of others, and the ability to celebrate their success have been pivotal skills that have served me well. Mastering the art of building meaningful and mutually beneficial relationships *is* my superpower.

As the daughter of Beloit, Wisconsin Police Department's first female detective, Officer Brenda Coleman, I attribute my success and innate leadership qualities to her guidance and influence. From my mother I gleaned not just a legacy of excellence but a profound set of values and principles that have shaped my approach to leadership and human interaction. I learned the importance of curiosity; I am relentless in asking probing questions and seek understanding, invaluable in dissecting complex business issues.

As a teenager, you could only imagine what happened when I strayed from the path for which I was destined; the thought-provoking and self-reflective conversations we had led to increased confidence and acceptance of my authentic self. During my high school years, I maintained a selective approach to my social circle, a necessity for maintaining peace within our home. There were clear boundaries on where I could go and with whom I could spend time—no questions asked.

One Friday evening, I made a spontaneous decision to attend a party. Initially
the party was chill, but the scene soon shifted to drinking and weed smoking. Wisely, I left the party, opting instead to cruise around the town, grabbed a bite to eat, and returned home earlier than planned. As events unfolded, the party was shut down by the cops. Mom knew I might be at the party, and she did not appear on the scene to avoid potentially embarrassing me.

Later that evening, Officer Coleman (Mom) returned home in her full uniform, woke me up, and engaged me in a heartfelt conversation. Though it felt like an interrogation, her approach exuded warmth and genuine care. As in an interrogation, she probed me with questions, though her approach exuded warmth and genuine concern. Her questions were aimed at imparting valuable lessons on decision-making, character development, and self-assurance. Patiently, she guided me in embracing my identity, my potential, and the importance of unapologetically accepting my authentic self. In that moment, her demonstration of grace marked the beginning of my journey towards extending grace to others and building relationships even in the face of adversity.

Mom also taught me the skill of reading the room and interpreting body language, which later helped me adapt and tailor my approach, engaging with others to achieve shared objectives. My mother could walk into a room and not only remember details of others' color, height, and location, she could look at a person and seemingly instantly intuit if they seemed to be of a nefarious mindset. I also developed an ability to quickly determine who is on board and who will need extra attention to accomplish the business objective at hand.

Most importantly, my mother emphasized the significance of expressing gratitude for the contributions of others towards the collective success of all (such as when she and her colleagues were solving a case.) This attitude of appreciation has helped me foster strong relationships with others, and continuously act in a manner that encourages and celebrates others. The alignment of my values to my actions has garnered unwavering support from colleagues. This has been instrumental in accomplishing our mutually beneficial goals.

From observing my mother's example, I've come to understand that success, particularly from a servant-leadership perspective, necessitates genuine care for the well-being and success of others. By prioritizing relationships and demonstrating a sincere interest in the personal and professional growth of those around me, I've been able to cultivate an environment in which everyone can excel.

~~~

My career began at the large consulting firms, where I learned the importance of humility, building rapport, mastering my domain, and mobilizing people to action. Whether interacting within the consulting firm, or with clients, I understood that by showing respect and appreciation for their time and expertise, I could count on a response when seeking assistance.

For me, this translates into a posture of humility, respecting others for their expertise and openly recognizing their brilliance—and then harnessing that brilliance to accomplish shared goals, whether it's implementing a complex system like SAP or rolling out a new process to complete a cyber assessment. Success in these endeavors is not only about achieving the goal but also about building sustainable relationships that ensure future collaboration and support.

Throughout my career, I have had the privilege of working alongside countless talented individuals whose work ethic and output were remarkable. However, I've realized that beyond mere talent and hard work, the ability to inspire and motivate others to act for a greater cause is a distinct skill not commonly taught or acknowledged. Effective communication is at the heart of building strong trusting relationships. Being honest, open, and timely with the acknowledgement of a job well done, and even injecting humor are essential components to recognizing the contribution of others. I believe listening intentionally to understand is key to engaging others and building trust.

Recognizing and attributing insights that others provide is a must in building trust and mutual respect, fostering an environment
~~~

where everyone can thrive and willingly share ideas. I bring a vision but how we achieve the vision will be the direct result of the great contributions of others. When ideas are shared and credit given to the person with the idea, to the team that took a concept and made it amazing, greatness is the ultimate outcome.

In the whirlwind of daily life, it's easy to overlook the importance of connecting with, recognizing, and celebrating others. I am constantly jumping from one project after another, shifting from one task to the next, often frustrated there just isn't enough time to get it all done. In these moments, I may forget to express a simple "thank you." Yet, I know that intentional recognition is vital for fostering a collaborative and supportive environment. So, I course-correct.

Reflecting on my experiences, I've found that remembering the impactful leaders and
environments where I thrived serves as a powerful reminder to acknowledge and celebrate the
contributions of others, like those leaders who inspired me to go the extra mile and who created environments in which I felt most valued and supported.

A recent instance when I could show appreciation to the team was during an enterprise migration to a new business continuity system. The migration was challenging, guidance changed daily, and as representatives to the organization, the team faced considerable resistance. The team not only provided valuable insights and suggestions for improvements, but also collaborated effectively to ensure the organization won. I demonstrated my gratitude by writing notes to their leaders and the CIO to update them on the program's progress; highlighting each team member's dedication and sacrifices for this initiative; and ensuring the central function heard and addressed their concerns. Most importantly. I publicly and privately thanked them with specific examples of how their efforts contributed to the organization's ultimate success.

I've learned by taking the time to acknowledge the hard work and dedication of an individual, I validate their efforts and inspire them to continue striving for excellence. Whether it's praising a team member for their innovative idea or highlighting the impact of a deliverable they've created, every act of recognition is a way of celebrating others' greatness and is foundational to building strong relationships.

~~~

Over 30 years ago, I pledged Delta Sigma Theta Sorority, Incorporated at Iowa State University, one of the best decisions of my life. At 20 years old, I bonded with six young women who became lifelong sisters. As many know, at the conclusion of the initiation process you are bestowed a line name and mine was True Inspiration—an honorable name, but I only focused on the length of the name, the hip factor or lack thereof, and most importantly how I couldn't afford to pay for the embroidery on a cardigan with all those letters. I was focused on all the wrong things! Fast forward to today and as I reflect on the sorority experience and my name, I clearly see my superpower was on full display. I encouraged and lifted my line sisters. I celebrated each of their wins that led to our ultimate success. Even back then I was connecting, building, encouraging and inspiring others to be their best.

I have celebrated and encouraged others all my life and those relationships have manifested into amazing experiences, and created a powerful network encourage me to strive to reach my fullest potential. As I have matured in my personal and professional journey, I now have a better appreciation of this superpower and I continue to use and refine it daily.

DaVon Gorman builds strong partnerships to lead change and execute organizational strategies. She is energized by and adds unique business value to companies undergoing large transformational change. As a results-oriented leader, her mission is to create excitement around a vision and enroll people on the journey to drive and implement transformational change.
~~~

Part II

Career-Building: Learning the Ropes

"If you think you are too small to make a difference, you haven't spent a night with a mosquito."

- West African proverb

Playing the Game: Early Lessons on Dealing with Office Politics

by TiJuan Lumpkin

Harmony in hierarchy, or the art of embracing positive office politics.

What comes to mind when I say *office politics*? Gossip, favoritism, backstabbing, manipulation, power struggles? When I started my first job after college, I was oblivious to office politics. I don't mean that I didn't see different stories playing out in front of me, some not too dissimilar to the soaps my grandmother watched religiously. But I didn't understand the meaning behind them or the underlying dynamics and how my ability to positively engage could influence the trajectory of my career.

Office politics are inevitable in any workplace, and they can have a significant impact on your professional success. However, not all office politics are negative or harmful. In fact, they can be a source of positive influence, collaboration, and career advancement if you know how to navigate with professionalism and positivity. I would say that by being more attentive to the dynamics in the workplace, I've developed some of my most rewarding work relationships.

Learning how to play the game helped me overcome my initial intimidation and develop skill and confidence in dealing with office politics.

One of my early jobs was at an automobile parts manufacturer. The plant was massive, and if memory serves, the building occupied over 1.5 million square feet. One of my responsibilities was maintaining the plant network infrastructure. Unlike a clean data center, the equipment out on the floor was not protected and was subjected to chemicals, grease, and various high temp areas. I would routinely find myself all over the building keeping things up and running. Shortly after starting there, I struck up a relationship with an older union electrician. I'm going to call him Paul. On the surface Paul and I had little in common. He was an older white guy from rural Indiana and spoke with a heavy Hoosier accent. Me, a young African American from the projects of Hamilton, Ohio. But Paul loved to tell a story and I enjoyed listening. We both enjoyed sports and telling jokes, and I think he liked having someone he could teach about all the inner working of the plant. Paul and I quickly became work buddies. One of the perks of Paul's job was that he had his own motorized cart for carrying his equipment all around the plant. The 1.5 million square-foot plant. The same plat that I needed to traverse to take care of my network equipment.

Our friendship soon turned into a beneficial partnership. As Paul and I would catch up on sports, tell jokes and connect, we would travel the immense manufacturing floor together so I could troubleshoot and repair networking equipment. Using Paul's electric cart vs walking or using a bike saved me hours of time every week, allowing me to get my work done far more efficiently. I wasn't scheming, being manipulative or playing *Game of Thrones*; I was just building a genuine work relationship that turned out to be mutually beneficial. But in hindsight I realize that was my first take at game of navigating office politics.

Let's look at how you can take a more intentional approach to mastering positive office politics like this, from day one.

First, you'll want to identify players and influencers in your organization and understand their goals, motivations, and preferences. By understanding who they are and what they want, you can tailor your communication and interaction with them to build trust, rapport, and respect, keeping your goals and actions in line with theirs to create mutual benefit. You may think this feels manipulative. But I ask you, how else can you build relationships if you're not interested in learning about the other person? You don't build up meaningful relationships in your personal life without learning about the other person and the same holds true in the workplace and can be done in a sincere and meaningful way.

We build trust and rapport with colleagues and managers by being authentic, helpful, and respectful. Trust and rapport are the foundation of any successful relationship, and they can enhance your collaboration, communication, and productivity. Treat everyone you encounter with dignity, courtesy, fairness. For me this means greeting the receptionist when I enter the building, catching up on the Pacers with Tim in security, and checking about her grandson with Cindy in maintenance. Everyone is important. Demonstrate that you value their opinions, ideas, and contributions. Offer your assistance, support, and feedback to others when they need it, and show that you appreciate their assistance, support, and feedback when it's offered. Be honest, transparent, and consistent in your words and actions, and know that it's okay to show your personality, strengths, and some vulnerability.

Be sure to communicate effectively and assertively by listening, asking questions, and expressing your opinions and ideas. Communication involves the way you convey your message, intent, and emotions to others, and how you receive their messages, intentions, and emotions. Effective and assertive communication can help you influence decisions, share information, and resolve conflicts. You'll want to listen, show interest and empathize with the speaker, and then ask questions, and express your opinions and ideas. Asking questions shows that you actively seek clarification, understanding, and feedback from the speaker, while expressing opinions and ideas means in a clear, confident, and respectful way demonstrates that you trust them with your thoughts.

Professional poise: Crafting a lasting impression in every workplace encounter.

To be clear, not all workplaces are great. I've been fairly blessed in my career to work in positive environments, which brings me to my next point: making sure that the culture and values of the organization where you work match your core values. If not, it's time to brush up your resume and find your next opportunity.

Finding an organization where you feel equally yoked allows you to align your actions and behaviors with the organizational culture and values. This helps you demonstrate your fit, commitment, and loyalty to the organization, and avoid conflicts, misunderstandings, and mistakes so you can perform better and deliver with minimum distractions.

As part of this goal, I've found it important to get feedback and advice from mentors and peers (ideally those who are well-connected) with experience and insight into the dynamics of a particular office. These well-connected folks helped me quickly adjust to new organizations/roles and point out any blind spots, allowing me to quickly begin contributing at a high level, and reducing the learning curve. Learning from their experiences, tips, and support, I was able to navigate office politics with professionalism and positivity.

While stories at the water cooler can be fun at times, I highly recommend avoiding gossip, rumors, and conflicts, becoming known instead as the person who does not partake. Otherwise, engaging in such negative office politics might damage not only your reputation and relationships, but also organizational culture and performance. Let's face it, you're going to hear a rumor from time to time, but that doesn't mean you have to spread it further. Be discreet, respectful, and constructive and let it end with you.

When conflicts arise, try not to take sides. When focusing on the conflict, you lose sight of priorities and the desired business outcomes. Instead, try to find ways to redirect to winning for the business and not just winning the argument. You then become known as someone who is constructive and focused on facts, solutions, and positive outcomes. When conflicts can't be resolved, look to bring in a trusted resource to help with mediation or resolution.

Let's get back to Paul, my unlikely workplace pal—someone with whom I would never have imagined building a relationship. Taking the time to get to know him meant creating a great partnership that made the workplace more enjoyable and provided additional benefits that helped me be more productive. Being open and respecting differences helped me foster a culture of inclusion and collaboration. With Paul and many others, this meant I was able to value and learn from their backgrounds, characteristics, perspectives, and preferences.

I hope this perspective on how to embrace office politics and play the game with intention is helpful when you're working in any corporate culture, allowing one more way to bring the best version of yourself to work.

TiJuan M. Lumpkin, Sr. is the Associate Vice President Collaboration and Mobility at Lilly, where he has a strategic role in managing key technology platforms and services that are essential for advancing the company's success in the digital age. Throughout his tenured career, TiJuan has held various roles across pharma.

Accidental Leader Meets Accidental Mentor

by Tim Hart

This is my story of believing I failed my way up into a leadership position and then realizing I was meant for it.

I'm an engineer and a systems architect with a good skill set and a decent mind for strategy. I've always been confident taking chances to increase the positive outcome of projects by automating my job. I constantly pushed the limits of my knowledge of other systems and connectivity to improve the efficiency of the platform I managed by myself. Our business used this system to price, pack, and ship medical supplies to millions of healthcare professionals and hospitals.

But I had a problem. The system was going down every three days, and everyone around me was a subject matter expert on my platform, but they could not understand how to fix the problem. Then, one day, I walked into work, and my leader presented me with a scenario*: The business loses $5 million dollars every minute this system goes down.* "I know you just started working here, but can you look at this with fresh eyes? This is what we hired you for. Can you fix it?" he asked.

I accepted the challenge and found issues within the first 20 minutes. Yet I needed to be more confident in my assessment because the business had already brought in high-priced contractors and solution architects. It couldn't be that simple. Surely the people who I thought were more highly skilled than me would've found something as easy as this.

Fast forward.

"Tim! Nice to meet you. This is the first time in six years we've made it a week without an outage on this platform. What did you do?"

I replied, "I simply followed best practice, starting with the fundamentals and customizing it to fit my situation."

The CTO and CIO were in shock, which turned into confusion, anger, and silence.

A few months went by, and I'd heard layoffs were coming. Was I worried? No, I was not. I continued to run my platform and train the incoming MSP to whom I would potentially transition my platform. I didn't get the email or tap on the shoulder and directions to go to the conference room.

After the layoffs, a leader appeared as I returned from lunch. "Can I speak with you?"

This is when I figured I must be in trouble and thanked my ancestors that I don't decorate my desk or customize my area to make it my second home. But it turned out he wanted to make me the Manager of Virtual Engineering due to how I handled adversity, my ability to think critically when everything was against me, and my ability to speak truth to power. I was in shock, not for getting promoted but more so because now I had to report to someone at the C-Suite level. I pondered what it would mean to be a manager and have a team to support me, to not have to do everything independently. Deal!

You might have thought it would then be easy for me. But I had been dealing with engineer and architect peers, not other leaders with differing motivations or priorities. At the time, there wasn't a mentor or significant family member in my life who could prepare me for this new responsibility of leading *people*. This was a journey I would have to navigate myself. First, I set a goal of learning the goals and motivations of each team member and the peers supporting my environment.

I arrived at work before my new team members, observing their behaviors and routines. I was determined to build a solid team to support me and put in the effort required to achieve my vision. My attention was drawn to two help desk workers in their 20s, who had been in the office before me all week, diligently clearing tickets and organizing their days before the rush.

Around lunchtime, I approached them and said, "I'm going to teach you guys how to run this environment better than me, so eventually, one of you will be my successor."

They looked at me in disbelief and skepticism. As I walked away, I heard one say, "Yeah, right, that's what they all say."

But within the first six months, they were better engineers than the current engineering teams. Within a year, they were both promoted to analysts, and within three years, they became senior engineers.

Despite facing leadership instability and incompatible visions, I remained focused on creating a solid team. I was determined to spend more time on strategy and less time doing tactical things that do not provide value to an enterprise company. My most significant requirement was building trust with my team so they would believe and understand that I supported them. With the team on board, we accomplished feats I never could've accomplished by myself, and that's when I first understood what it meant to be a good leader.

Within a year, and after various accomplishments, I was tapped on the shoulder to lead a global team focusing on stability for a Fortune 7 company. *What? This must be an accident. This can't happen again. Was it a setup?*

No, it was not. I was made director within one year of transitioning to manager, thanks to the team's efforts in building a coalition and advocating for doing things the right way instead of the easy way.

Once in the new role: organizational changes, hurricanes, wildfires. Whoa, this was not what I signed up for. Why does it seem like I'm putting out more fires here than when I was an engineer? The work started affecting my home life and relationships, and right at the moment when I thought about giving up, I met someone: a 69-year-old retired African American Air Force pilot, Jack Dawson Sr.

At first, he stared at me, and I stared at him. Then, I extended my hand and introduced myself and he did the same, asking this young brother to talk with him. It turns out this man had been one of my NCOs while I was in the military. He remembered me from all those years ago, from Private to Senior NCO, and now offered to be my mentor now because he'd been watching me since he'd moved into the neighborhood.

I had never had a mentor, never looked for one, and never saw the importance of it. I conditioned myself always to think that I would build myself without anyone else's support. But once I had someone to talk to, who understood my struggles as a Black man, listening to him talk about leadership and career situations he'd experienced, I was able to translate those lessons into my life. One thing that stood out to me was that early on he realized he had the power to control his destiny and the narratives his peers and leaders applied to him.

As I applied the knowledge gleaned from my mentor to my career, I then opened doors for others who wouldn't think they had the skillset to work in this field, generating opportunities. The hurricanes began to recede, and soon accolades started rolling in—as a result of building a coalition within the business based on the knowledge I gained from my mentor. Later, I saw that all it took was one person supporting me at every career level who believed I could succeed. Sometimes, it's not enough just to work hard.

Timothy Hart is the Director of Engineering at McKesson Corporation. With a sharp eye for innovation and a strategic mindset, Timothy is a dynamic and results-driven IT professional who has excelled in driving impactful technology initiatives, aligning technology solutions with business objectives, and delivering results. Timothy is responsible for overseeing all virtual technology of the company's IT I nfrastructure, systems, and operations.

The Art of Listening, and Other Gems That Made Me Successful

by Floyd Smith

Success can't be turned on like a light switch. The sum of actions, events, and fuel success. My own road to success has been filled with curves, bumps, and hills, but I continued until I reached smooth ground. When I think about how I got here, several key attributes come to mind, professional and personal.

Active listening matters.

Early in my career I failed miserably at active listening as I was often focused on responding immediately instead of concentrating on what others were saying. During one weekly touch-base meeting with my manager, I nodded while she covered key strategic information, mentally focusing on my response, so determined to voice my opinion that I failed to listen or take in what was being shared. This action resulted in me delivering 75 percent of the project scope. Upon completion, I expected to receive praise for completing my assignment ahead of schedule, but instead, I was questioned for not delivering 100 percent.

I explained my understanding of the project scope and quickly determined what details were missed, a humbling action I will never forget. We were eventually able to complete the remaining 25 percent after a two-week delay from our original end date. As a leader I owned the oversight, and I learned a valuable lesson.

Active listening is an art that comes naturally for some leaders but for others requires continuous practice. The goal is to listen to someone while focusing your attention on what they are communicating, not on what you will say when it's your turn to speak. You want the speaker to feel heard, respected, and understood. And you want to be sure you get what they say.

While managing a system implementation project for a large national health system, I was called to my manager's office to review the project status. I felt nervous as I walked to her office, as this customer was demanding and hard to work with. Our project had started slow however, we'd gained momentum and were now on track. My manager opened the meeting by highlighting project challenges faced at onset. Defensive, I immediately interrupted to explain what happened and that we are now on schedule.

My manager paused and said, "Relax. I'm not here to shame you. Please hear me out." She then commended me for building a relationship with this difficult customer and as a result, the customer has requested that they only want me managing their projects. This exchange taught me a valuable lesson about listening (as well as not taking things personally and becoming defensive).

Over the years, from that experience and many others, I've learned a few things that help me with action listening: When interacting with someone I look directly at the other person, putting aside all distractions (including cell phone), and give my full attention to the individual. I engage with the use of positive body gestures or nodding occasionally to acknowledge understanding.

Follow up or follow through.

While working in healthcare as an implementation specialist, I often arrived in the office earlier than most, utilizing the quiet atmosphere to gather my thoughts, organize my calendar, and get a head start on the day. Many days, when visiting our location, our general manager would stop by my desk in those early minutes to chat, conversations that seemed quite casual, getting to know each other on a first name basis. I thought nothing of the interactions, but on one of his visits, towards the end of our chat, he asked me to follow up with him before the end of the week to discuss an idea. I promised I would.

After completing my to-do list that very day, I swung by his office, where he welcomed me in. I had no idea what was in store for me or my future. I was intrigued to hear of his initiative to implement formal project management doctrines/principles for all enterprises projects and initiatives moving forward. That was followed by, "I would like to send you to Georgia Tech University to attend a 10-week Project Management Certification course." I was surprised by the offer and immediately accepted.

Early on I had had no idea that our morning chats were more than trifling conversations. But they led to obtaining my Project Management Certification, a pivotal point in my career, opening up previously unavailable employment options.

Two things happened here: I listened actively, and then actively demonstrated my commitment to follow up by honoring a promise. This shows that you're respectful and trustworthy. Follow up costs virtually nothing but time and effort, but it can lead to unexpected opportunities.

Self-confidence is yours if you decide it is.

In my career I've struggled with self-doubt. While professionally, I knew I was competent and had command of my area of expertise, but I struggled with shaky confidence. When I shared my battle with my mentor, he replied, "What in the hell is this self-confidence bulls%$& you're talking about? I've known you for a while now and have witnessed your growth as a man and leader. What are you afraid of? If it's failure, get over it. People fail all the time. It's the lessons learned from our failures and the actions taken to prevent them from re-occurring that are important. Floyd, *never* doubt your abilities."

By listening actively, I took in his words, and decided to believe him. From that day on, I've never doubted myself or worried about self-confidence. I'm often the only person of color in most of my meetings at work. Each day now, in every situation, I hold my head high and exude confidence.

Believing in yourself is a struggle for many African American leaders at various points throughout their career. We have the education, skills, and experience needed to effectively perform (and often outperform) at our jobs, but a lack of self-confidence can prevent us from progressing to that next level. This is often a variation of what is referred to as "impostor syndrome" where individuals second guess their right to their position.

Self-confidence begins with believing in yourself and in your ability to accomplish what you want in life. Knowing your strengths and weaknesses can impart a certain swag and allow you to take full control of your life. Some people might mislabel swag as cockiness; however, I think of it as self-confidence. And it's critical.

Ask for help without hesitation.

Pride often stands in the way of becoming successful; worrying about what others think will only distract you from clear and logical thinking. Pride also can keep you from seeking the kind of help that matters because you think it may make you look weak. This faulty reasoning kept me attempting to solve problems that I did not understand instead of asking for help.

Every successful individual has received guidance or counsel from others throughout their career. I've sought counsel from my mentor for so many issues. Early in our mentor/mentee relationship, he told me he would be by my side no matter what. Good or bad, day or night, do not hesitate to reach out. He shared with me that he too learned a great deal from his mentor, and we must continue the practice of paying it forward. Through his counsel, when evaluating job offers, I've been able to negotiate better compensation packages, or leverage his network for assistance. In many cases, I've been able to return the favor. Even now, as a successful leader, I still regularly seek counsel from mentors and trusted colleagues. This does not mean I'm incompetent. This means I'm able to recognize and embrace when I need assistance from others, and when I can help others equally.

Building relationships is paramount.

My success has often hinged on my ability to form strong, dynamic relationships built on mutual understanding, respect, and trust. Developing and maintaining long-term relationships is probably the most critical skill I practice daily. Relationships provide access to key people and vital information, and exposure to what I can't access on my own. Once relationships are established, maintaining them is crucial, and can include anything from having coffee, attending social events, or hosting dinner.

When I was enrolling in college, my mother and I were greeted by a gentleman in the registrar's office asking if we needed any help. He provided directions to our next destination and told me his name again and where he could be found if I ever needed anything. From that day forward, Mr. James Wallace was my mentor, until he was called home on November 28, 2018. The relationship I forged with Mr. Wallace changed my life; he exposed me to things I never knew existed and provided a lifelong lesson in the value of maintaining connection.

Staying open to feedback.

Over time and experience, I've learned that being open minded and receptive to feedback has been a major contributor to my professional development, personal growth, improved decision making, and relationship building. Hearing what you need to improve doesn't always feel good. But you need to listen anyway. When receiving feedback, practice active listening; your goal is to hear and understand, and learn something about yourself, not to dispute and defend.

My first job out of college I worked in public health where documentation is critical. One day, my manager handed me a report that I'd submitted a week before; it was now full of red marks. She sat down next to me and walked me through the entire report summarizing her comments and suggestions.

This was not fun to hear, but I thanked her for providing open and candid feedback. Then I took time to process the feedback and understand why it was merited. In my effort to complete assignments in a timely fashion, I had failed to pay attention to detail and proofread my work. I learned that quality was just as, if not more important than timely completion. One of the biggest mistakes professionals can make is doing or saying nothing after feedback is provided. So, I closed the loop by following up with my manager to share the actions I planned to take based on her feedback. And then, I followed through.

Feedback should never be taken as a personal attack, but an opportunity to find areas of growth or opportunity. When I allow myself to think of feedback as constructive, given with positive intent, the benefits are greater for all parties.

Success doesn't happen by flipping a switch. Rather, success blossoms with consistent attention, like a plant in a garden. It's the slow, deliberate application of purposeful attention, commitment, and actions that allow the growth that leads to success.

Floyd W. Smith II is an Information Technology executive with 20 years of experience in cybersecurity, leadership, systems engineering & operations, and compliance. He's helped fortune 10 companies streamline business operations, improve cybersecurity, and increase productivity. Floyd has a passion for mentoring students, developing young professionals, and volunteering in his community.

Amplifying Your Voice: Training the Workplace to Tell *Your* Story

by Spencer E. Rollins

I have had the pleasure of working for great leaders who have been critical and beneficial to my career growth and development. Likewise, I have also worked for and observed horrible leaders who did not have the best intentions for others' careers.

A mentor during my early stages of middle management once said, "You can destroy a person's career passively and never do anything overtly against them." The mentor was in the C-Suite and showed me firsthand how an upcoming leader, targeted to join the C-Suite, was sabotaged.

Before I get too far into this example, it should be known that the C-Suite is like a fraternity/sorority. It can be very cutthroat to be elevated to and remain at this level, and getting there is often dependent on someone already in the C-Suite advocating for you.

Back to the story. The C-Suite executive explained to me, "If you continually tie someone's name to failed events or projects, regardless of the fact the person had anything to do with them, it will leave the audience with a subconscious perception of failure associated with the targeted individual." Then I saw him put this terrible tactic into action.

During a meeting, the leadership team was speaking about a project we will call River Run, which was over budget and did not hit the target timeline. I then heard my mentor state, "I think that project was run by Jane Doe." He knew that the project was not run by that individual, but with no one in the room to correct him, the seed was planted.

I saw this tactic of sewing the seed of failure associated with Jane Does repeat several times. The plan worked flawlessly and a talented individual's career never advanced due to unconscious perception of failure purposely planted, casting unfounded doubt within the executive leadership fraternity. Jane Doe was deemed unworthy because of an incorrect perception of not being able to deliver.

While this mentor set a poor example in this one area, he also taught me early in my leadership journey that managing my own perception among others, and amplifying my own voice were critical components to success.

Another example of an experience with a different poor leader: During my exit interview, I said, "I appreciate everything you taught me." He looked taken back, probably because he knew he had been quite busy taking credit for my work and undermining my success. Based on our history, he should have sensed the sarcasm, but he pulled his shoulders back and smiled as if I meant what I said as a compliment. He was a good example of why it's a bad idea to undermine a team member and take credit for others' efforts.

A great quote I love, from Australian author Charlotte Freeman goes like this: "Every experience, both good and the bad, will eventually get you to where you are supposed to be. Every win, every loss, and every mistake." Something really *can* be learned from every situation. I try not to wallow too much on negative experiences, which we will all have, but focus on the lessons that can be learned.

I was recently asked a very interesting question: Which is more detrimental to your career? A person of influence who will do nothing to help you advance? Or someone that you know is standing as an adversary towards your advancement?

My answer was a person who can help but will do nothing. But for combatting either, it is critical to network, to build a wider community of supporters outside of your direct leadership where you can create your own unfiltered voice.

How to develop your own narrative.

How do you create that unfiltered voice and be heard? I will share some tools that have been very beneficial to me in multiple organizations—and once helped save my job.

- Annually, I develop a roadmap tied to corporate objectives plus those I have created for teams I lead.

- I discuss this roadmap with my leadership to validate my thoughts are in line with theirs.
- Next, I develop detailed team goals to meet the objectives in the roadmap. The goals are then cascaded down to team members so that everyone has a vested interest in achieving the goals. For example, if the company had a corporate initiative for Cloud use, I may have a cascading goal for team members to take and/or lead Cloud training, obtain certifications, identify systems that are candidates to move to the Cloud, etc.

Here is the part that is critical to using this tool to create your own voice within the company. I create a monthly organization dashboard. That dashboard ties to key performance indicators, defined goals, and the organization's roadmap. The key to the monthly dashboard is ensuring that the reader has a clear understanding about how my activities are affecting the company's bottom line (e.g., revenue, expenses, risk, etc.).

I also list problems or key decisions that need to be quickly addressed. This step helped save my job once. Within the dashboard I documented where I communicated a significant business risk that needed to be mitigated. I offered multiple options and showed that I was denied funding and resources. This lack of judgment by my leadership resulted in the company experiencing a significant loss associated with the predicted threat.

In another instance, I was responsible for rebuilding a cyber security program after several companies merged. The executive who recruited me to lead this effort did not stay at this company very long, which should have been my first red flag. I quickly discovered the executives I reported to were very ineffective and led through fear tactics (replace people and not address the problem). Because I knew the dangers of being associated with that type of ineffectiveness and approach related to enterprise risk, I began documenting my teams' activities and challenges in a monthly dashboard.

Surprise, surprise. The company was impacted by an avoidable security incident. The fact that I had highlighted the significant risk to the company and noted the needed remediation steps in a dashboard seen by the Chief Information Officer, Chief Compliance Officer, and Chief Legal Officer, I maintained my job after avoidable loss to the company. Because I had relationships outside of my reporting line and other executive leaders had seen my monthly dashboard, my employment continued. (And the company decided to part ways with the C-Suite executive who had opted not to prioritize the warnings that were clearly given to him in black and white or better yet PowerPoint.)

While my dashboard tactic could be considered a CYA move, it's also a not-so-subtle way of coaching others on what to say about you when you are not present. This experience taught me to be my own pilot and not rely on wishing for the best and hoping my leadership would say and do the right things when I'm not around. I learned to drive the message I want, the way I want to it told.

People are easily influenced, positively or negatively, when information (stories) are printed, posted, and promoted. Marketing companies spend millions of dollars on social media to subconsciously sway opinions. The dashboard is one way I can amplify my voice within the company using a similar strategy; the monthly dashboards highlight wins that can be appreciated by any executive and also clearly shows risks. These dashboards usually leave the readers with the perception that I am a strategic thinker and game changer that positively affects the company's bottom line. This approach also helped propel my career by letting leadership quickly see the value my teams produce.

The key to developing an effective dashboard is quantifying key performance indicators to the business. Tying performance indicators to one or more of three key business drivers will quickly be recognized and appreciated by anyone in the C-Suite: increasing revenue; decreasing expense; risk mitigation.

I strategically use the dashboards as a tool to help tell my story. People often remember what they've read and will often repeat data just as it was presented. You give people the language that you want them to use, to minimize the risk of them saying something that you don't want them to say.

This tool has significantly amplified and highlighted my voice in rooms where I was not present and created opportunities for me to start being invited into other rooms, after someone recalled and mentioned data they had seen in my dashboards. As a result, I have been given increased responsibilities and promotions to duplicate this pattern of success in other areas of the business.

Finding creative ways to tell your story (and sometimes CYA) and have your story told by others the way you want it told can lead to advancement. Either you tell your story or have it told for you. Using a baseball analogy, building a strong network at multiple levels within the company is akin to making contact with the ball. Creating a tool that amplifies your voice and sharing your story as you want it told is like knocking the ball out of the park.

Spencer Rollins is a business leader with 20 years of international cybersecurity and risk management experience. He has an extensive advisory background leading CIOs/CISOs to lower enterprise risk to meet company objectives and/or regulatory requirements. Spencer has also built cyber risk organizations from infancy and bare structure to fully functioning, robust programs.

Embracing Hard Lessons on the Path to Executive Success

by Donald Graham

When I joined the United States Marine Corps, I was a young Black man, wet behind the ears, fresh off my high school valedictory speech who thought he knew it all—green, highly motivated, salty, ready to serve. My time serving this country taught me a lot about myself, a lot about how to be political/strategic, and a lot about being a leader. But my career, both in the military and corporate America, has been rife with examples of me learning the hard way.

Building: the importance of a network.
It was 1983, we mustered at attention at 0530 hours on the airport runway aka the flight line, eyes red, anxious but ready for deployment, young and excited about going overseas as an expeditionary force. After waiting four hours, a UH-1 helicopter picked us up at El Toro (CA) Air Station, and on the trip to Alaska we drank and got rowdy until the sergeant major cut us off. The next leg of the trip from Alaska to Tokyo was mind-numbingly boring, 13 hours of hangovers and fear of the unknown. If wasn't long after I found myself stationed in my first overseas post, Okinawa, Japan as a young Lance Corporal (E-3), a survival equipment specialist (military occupational specialty 6060).

I was lean and green, but incredibly hardworking (and back then I thought hard work alone was enough), so when my gunnery sergeant told me, "I'm gonna put you up for Marine of the Quarter," I was ecstatic. Taking those Marine Corps extension courses was finally paying off. I threw myself into preparations. *What does it take to be Marine of the Quarter? How should I study? Who will be on the panel? Who should I talk to?* As I thought through possible scenarios and questions that stood between me and the award, I felt the weight of the nomination and realized that I was alone. I was on an island surrounded by other Marines, but I didn't have a network I could turn to for guidance, answers, or even reassurance.

The day of the selection panel, I nervously slid into my dress uniform wiping my clammy hands on the front of my freshly creased pants. I was clean (as clean as the board of health) and confident in my accomplishments, but nervous, very nervous. The board, comprised of senior enlistees, asked many questions and as I fielded all of them with ease my confidence grew, chest out, making direct eye contact.

The final question sounded like what they thought would be a curve ball: "Who is the president of Libya?"

I answered in my most enthusiastic voice, "Colonel Kaddafi!"

They all looked at me in surprise and then promptly thanked me for my time with a curt, "We'll get back to you.

I walked out feeling like I had nailed it, that I had it in the bag. So, imagine my surprise when my Gunnery Sergeant later told me, "You didn't get it." With no preamble and no qualifiers, he continued "The guy who got it had connections. You were the best candidate, but nobody really knew you." I sat there with my jaw on the floor as he patted me on the back and sent me on my way with "maybe next time."

There are a lot of moments from my military experience that have shaped me into the man that I have become—two tours overseas, directing scores of Marines in times of crisis— but this has remained a defining moment for me in understanding the importance of networking. The guy who got the title Marine of the Quarter didn't outwork me, and he didn't answer questions better. But he did know the first sergeant and sergeant major. He had spent time learning the key players, influencing the decision makers, and he understood the game in a way that I didn't at that time. Hard work alone is not enough. You need a network of advocates and sponsors. You need to get known.

Effective leaders hire great leaders.

I have held many jobs (not all glamorous) and I have been lucky enough to be in the right place, at the right time, to make some valuable connections. While working, mostly volunteering, at my friend's auto shop, I met local movers and shakers, including a local politician who ultimately offered me a job as a legislative aide to two council members. The personal growth and business experience that I gained in that job is something I continue to leverage today.

While working as a legislative aide, I got a call to help straighten out the municipal golf course which was in disarray. The course sat atop an aquifer (the water source for a city 15 miles away) and was a protected area, but revenue for the course was down to less than $250,000 per year and it was losing money. I was tapped to help pull this cash-strapped operation together.

Upon my first visit I tugged on the collar of my heavy overcoat against the wind chill and looked out at the course in disgust—it was in shambles! Yes, it was wintertime, but I noticed virtually no employees on site. The place looked abandoned and unkempt, with golf carts haphazardly parked, broken equipment littering the landscape, and only the cashiers as gatekeepers. *Shouldn't there be maintenance work happening in the off season?* The wild deer population were frolicking all over the golf course. I took in my surroundings and the task before me.

When I could gather the employees, a senior worker relayed that, "During the winter we come in, but there are no head groundskeepers, so we don't know what to do. Plus, nobody plays golf here in the winter so there's nothing *to* do and we just hang out till it's time to go." *So, the taxpayers are shelling out hundreds of thousands of dollars paying your salaries and you're not doing any work?* I made it my mission to fix this. In every opportunity, in every job, there is an opportunity to add value. In this instance, while there was snow on the ground, there was still a lot of work to be done: machine repair, building maintenance, process improvement, etc. The key would be having someone in place to drive the work; I needed a groundskeeper.

With the help of a mentor, I did some research and got in contact with an experienced groundskeeper at Baltusrol, a private golf club in Springfield, New Jersey. He connected me to a solid candidate whom I hired for the head groundskeeper role. Interviewing groundskeepers and making an offer was my first real hiring experience, but it turned out to be a good one. The head groundskeeper I hired was competent and turned around the operation within a year. With his leadership, the team fought off cinch bugs (a major pest for golf courses), improved processes, and turned the course into a well-known destination for golf enthusiasts. What was previously a money pit turned into a course where customers would happily pay a respectable fee for 18 holes of golf. I showed up at the course to personally monitor, usually at the 9th hole receipt check, and observe the business (pilfering was a major problem), but I had learned a valuable lesson in execution: hiring the right leaders who can enforce accountability.

Two years later, the municipal course was running like a well-oiled machine, membership was up, and revenue had improved by over $400,000. My time at the golf course feels like a lifetime ago, but the growth from that experience continues to serve as the foundation for how I interview and hire great leaders.

Learning the Hard Way: Or, as Yogi Berra once said, "It ain't over till it's over."

Early in my career, I was up for a job and flew to Atlanta for the final interview. On the plane I mentally reviewed my conversations with my sponsor, then relaxed and closed my eyes, confident that I was prepared and the right person for the job. We touched down early, another good sign, and I made my way to the nearest Waffle House for breakfast. Later that day, I drove to the corporate office, swiped my badge, waved to security, and boarded the elevator to go secure the job that I knew was already mine. Sure, I was nervous—this could be a transformational moment of my career—but beneath the nerves lay a grounded sense of self-assurance. This job was mine.

Go time. I looked down at my watch at 5:30 pm as I entered the interviewer's office. As I expected this interview to be over quickly, I had booked a 7:30 pm return flight back home to Jacksonville, Florida. (If you know anything about Atlanta traffic, you know that I left no room for delays. The point is, I was more worried about making my flight home than the interview itself.) My interview was with the regional manager, who had a significant remit (approximately $150 million of software and hardware sales quota);

he began with a few softball questions: *What do you think about the role? Why are you interested in working here?*

I had been thinking about this role for a long time, so I was prepared with well-crafted answers. When (not if) I got this job, I would be a senior manager for the southeast United States, covering Tennessee, Florida, Georgia, and North Carolina. It was the natural next step in my career progression, and as the interview went on, I felt more comfortable and confident. But at the 45-minute mark, I glanced at my watch. More than once. The interviewer, noticing, asked, "Hey, can you stay? This might go on a little bit longer."

I feigned regret and told him, "I'm booked on the 7:30 Delta flight back to Jacksonville. With traffic and airport lines, I really need to leave in the next 30 minutes."

We wound up the interview and I shook his hand with a sense of entitlement. I made my flight, and I was home on the couch watching TV by 11:00. Only briefly did I allow myself to wonder, *Did I blow the interview? Maybe I should've stayed longer?* But those thoughts came and went with no real self-reflection.

My main competitor was a military academy graduate local to Atlanta, but in my mind, I was still the better candidate; he had less experience and was new to the region. But reminiscent of my Marine of the Quarter experience, I'd learn later that he was working all angles and positioning himself as the better choice. He had built a strong network in the local office, had daily contact with the hiring manager, and he used his unlimited time during the interview to lobby for the position while burnishing his military credentials.

About a week later, I received a call telling me the role went to the less experienced, ex-military candidate. I was devastated. Scratch that, I was pissed! I was the best person for the job, and I immediately requested a meeting with the engineering director to discuss the decision. Two days later we had a heart-to-heart conversation that would change my professional life.

I logged into the video conference and adjusted my suit jacket to make sure I was poised and ready. Shortly into the call, the engineering director candidly told me, "You were outmaneuvered. He got the job because he worked all the angles." He looked in the camera and delivered four words that would continue to haunt me, “Better luck next time.”

In subsequent interviews, gone was my casual “I got this in the bag” approach. I showed up and played to win. Six months later, I was promoted to senior manager with responsibility for close to $300 million worth of business.

My parents have often told me that I like to learn things the hard way. And that used to be true for me. But now I know that a large part of being successful is understanding the game. Every play matters. Lesson learned.

Donald Graham is Global Director of ybersecurity at Deloitte. He has 25 years of experience leading high performing teams and delivering technology solutions in multiple domains. A trusted advisor and mentor, Donald is passionate about STEM education and electrification.

Part III

Transition to Senior Management: Climbing the Ladder

"No person is born great. Great people become great when others are sleeping."

- African proverb

From Doubt to Destiny: Orchestrating My Transition to Tech Leadership

by Parisia G. Clyburn-Patton

Amplifying my voice, from silence to symphony.
The fluorescent lights of the senior leadership meeting seemed to hum in an unfamiliar frequency, highlighting the gulf between my present and a memory that surfaced. I sat, shoulder-to-shoulder with seasoned industry veterans, their words flowing confidently, referencing intricate concepts and historical insights. Picture this: in between the high-stakes discussions and strategic maneuvers of the executive realm, there's an intermission of lively banter and camaraderie. Amidst these discussions, executives swap stories of their exploits representing the company and encounters with industry big shots. Let’s not forget the fancy dinners—at times, nine courses of culinary delight served up by chefs who could make Gordon Ramsay blush. Along with the clinking of silverware, these execs share nods of approval, not just for the flavors, but also for the virtuosic achievements of one another’s offspring. It's a scene where sophistication meets humble pride, where the clatter of cutlery harmonizes with the laughter of peers in a delightful sonata of success. Just another movement in the grand opera of executive life, seasoned with a dash of dry wit and bonhomie.

But let's be honest, folks. Stepping into a room full of heavy hitters can feel like wandering into a symphony orchestra mid-performance, clutching a kazoo. You earned your seat, you know the music, but that tiny instrument just doesn't scream authority, right? It was this stark juxtaposition that struck me first—the awe of being conjointly amidst “intellects and innovators," and simultaneously the "mental bars of restraint" threatening to suppress my voice. There I was, a Senior Vice President (SVP) in a global financial powerhouse, co-leading a massive transformation program, but still facing the familiar foe of self-doubt disguised as an internal orchestra of "maybe-I'm-not-enough" violins, weaving whispers of inadequacy around my rightful seat at the table.

This wasn't a new melody. This internal orchestra, conducted by self-sabotage, had played its discordant tunes on various stages throughout my professional life. My mind drifted back to a different kind of learning arena: the worn wooden kitchen table and antique den chairs in my childhood home. My father, a man who believed knowledge held the key to unlocking any door, insisted on thorough research for even the most casual family discussions. Forget quick Google searches; our arsenal was the hefty *Encyclopedia Britannica* collection, each volume a portal to hidden worlds. From dissecting a historical event sparked by a dinner-table debate to tracing the etymology of an unfamiliar word, our days turned into mini-expeditions, my sister and I constantly trading roles as we shifted between playing the designated "investigator," meticulously combing through pages, or the "interpreter," weaving information into cohesive narratives for debate.

In my teenage rebellion, I saw these sessions as intellectual boot camp, to be avoided at all costs. When it couldn't, I sought shortcuts, skimming pages instead of savoring the richness of information; now, sitting amidst these leaders, I felt a pang of regret. Those "tortures" were opportunities I'd underestimated. They cultivated an insatiable curiosity, a love for deep dives, and the ability to weave diverse perspectives into logical notes. While others casually cited historical precedents and industry trends, I felt a self-imposed silence settle over me. Had my early shortcuts left me ill-equipped for this level of discourse? My father's voice, echoing in the recesses of my mind, offered a stern yet encouraging nudge. "Knowledge is power," he used to say, adding, "The fruit does not grow on the trunk of the tree; you must go out on a limb to pick." He was right. While shortcuts had given me the illusion of understanding, they hadn't provided the substance I needed. This realization became a turning point.

Embracing the lessons of the past, now in a professional capacity, I embarked on a personal learning renaissance. No shortcuts taken! I devoured industry reports, engaged in in-depth conversations with my peers, and sought out mentorship from seasoned leaders. Slowly, the gaps in my knowledge filled and the self-imposed silence began to dissipate.

I found self-doubt too wasn't an insurmountable foe. Mindfulness practices helped me recognize its deceptive whispers and replace them with affirmations of self-worth and accomplishment. Celebrating successes, big and small, strengthened my belief in my own abilities. Initially, self-doubt, the persistent little gremlin, wouldn't be silenced, but mindfulness became my shield, and stepping outside my comfort zone my new mantra. Collaborating on side initiatives that aligned with overall strategies allowed me to showcase my expertise and leadership capabilities in new contexts. Embracing calculated risks, even when met with uncertainty, demonstrated my initiative and commitment to driving progress. With practice, I silenced the orchestra of self-doubt and amplified my own voice, conducting my own symphony.

My voice, initially hesitant, gained confidence as I shared insights gleaned from my newfound depths. The journey from silence to symphony was marked by both challenges and revelations, yet eventually transformative. Sitting in leadership meetings now, I confidently dive into conversations. I reflect on my early years around the kitchen table now and realize that those childhood research sessions instilled a hunger for knowledge, a skill for analysis, and the ability to connect seemingly disparate dots.

Trusting the climb from doer to strategist.

The ascent to senior leadership felt like traversing a vast desert. As a mid-level manager, I was the dependable one, the get-it-done warrior who dove into the trenches alongside my team. Now, surveying the sprawling landscape from this new vantage point, I saw a different kind of leadership was demanded. Picture me, a bull in a China shop, used to charging in and rearranging the shelves myself. Now, surrounded by delicate figurines and intricate blueprints, the urge to "fix" was overwhelming. But those figurines were my team, and the blueprints, our collective vision. Letting go of the doer mentality proved harder than I imagined. Was I truly equipped to navigate this expansive new terrain? Doubts about my strategic compass gnawed at me. Yet slowly, confidence bloomed, allowing me to trust myself as a strategist.

Learning to lead was more than wielding authority—it was about guiding the hands that did the work. This transition came sharply into focus during a pivotal moment. Entrusted with 200-plus quality organizations for a global transformation program, I encountered a manager's stark candor upon her return from London, preceding my visit. "They don't want you there. Go fix it," she declared, leaving me momentarily stunned. Yet I discerned the trust and confidence underlying her words. Armed with the newfound perspective, I arrived in London, ready to strategize and course-correct. Over the next two years, I forged connections with key stakeholders, steering our collective efforts towards strategic success. Looking back on that defining moment, I am reminded of an invaluable lesson—true leadership transcends mere authority; it lies in the ability to inspire, to empower, and to chart a course towards a shared vision. I still carry with me the wisdom gained from that humbling experience—from doer to strategist, from follower to leader.

Balancing the new senior management landscape.

Imagine learning how to steer a ship in real-time in the midst of a sudden storm. That's what transitioning to a senior technology leadership role felt like, in the beginning. Emails, meetings, and responsibilities surged, threatening to sink my carefully balanced life. But even amid such chaos, the commitment to family remaining a top priority was met with some level of ease. Basketball games, swimming meets, and even the sports events of the non-profit organization my husband and I shared. I don't think I missed a beat.

I believe everyone saw, firsthand, the seriousness I brought to my new role, and the respect I held for its challenges. But they also witnessed how I orchestrated my life, setting boundaries and carving out moments for family and personal rejuvenation. Each shared meal, family meeting, or Sunday park outing became a refueling station, reminding me why I embarked on this leadership journey—not just for personal achievement, but for the shared journey we'd all take together.

Open, clear, and frequent communication was the cornerstone of successful initiatives and became my secret weapon with my team. It wasn't about scaling a mountain top or surviving the scorching sun, but navigating a complex landscape, setting clear expectations, and finding harmony between professional and personal commitments. This newfound balance was aimed at thriving, ensuring I led with clarity and compassion, not exhaustion. This journey wasn't a solo act, but a symphony. My sons, by witnessing my dedication to both work and family, learned valuable lessons about responsibility, balance, and the power of commitment. And my team, empowered by clear communication and realistic expectations, thrived alongside me.

Finding my crew and dealing with code-switching.
This senior management gig ain't a solo act. Allies became my cheerleaders, and I nurtured those crucial relationships. As my father said, "There's a reason why a path exists." The path to leadership is rarely solitary. Allies saw my potential, believed in my voice, and amplified it whenever possible. Seeking sponsorship, a powerful force in career advancement, was equally important in my development. These mentors provided guidance, opened doors, and advocated for my ideas, helping me navigate the sometimes-opaque systems of power.

Remember "chameleon-ing" in school? The juxtaposition dance that crept up (without me noticing at first), was a disguise that compelled me to shift from authenticity in both my personal and professional lives yielding, at best, a constant negotiation of identity. In professional spaces, there's often a pressure to conform, to mold ourselves into a version deemed acceptable. But I came to realize that authenticity is the real power, and suppressing it only weakens our voice. So, I made a conscious decision to integrate my heritage and experiences into my leadership style, weaving them seamlessly into my communication.

Drawing on personal anecdotes and historical context, I began to speak in a language that resonated deeply with me. I shared stories of my paternal grandmother, a Black entrepreneur navigating the challenges of the 1940s through the early 1970s. My father's reminiscences about her, and my own reflections on young Black and brown women in the technology sector, added layers of depth to my narrative. My grandmother's journey, from humble beginnings to empowering other Black girls to become educated professionals, became a powerful symbol of resilience and determination. These were expressions of empowerment. Turns out, authenticity rocks.

The power of shared experiences.
"Leaders tell stories," Dr. James Rodgers (ITSMF Executive Academy Director) often reminds us. Though unsure initially, I embraced sharing and found strength in community, because as it turns out, I wasn't alone. Connecting with other Black and female leaders was a revelation. Their stories resonated, their resilience empowered me to push further, and together, we advocated for others like us. This was important and allowed me to expand my network, and more importantly, breathe. Knowing I wasn't alone was a source of immense strength. Surprisingly though, the relationship I started to cultivate the most was not with other Black and female leaders, but with a white male, a senior executive leader whose son was experiencing challenges in a situation resembling that of my youngest son. As a result of our connection, I was invited to meetings, forums, and leadership events, of which I had no prior knowledge. This, in turn, led me to connect with two additional white female executives, forming personal and professional relationships, leading to invitation-only executive vision and strategy discussions. Connectedness fueled my growth as a leader, opening doors and fostering advocacy. Sharing strengthened my resolve.

Conquering the symphony within.
The journey wasn't easy, but the rewards have been profound. Overcoming self-sabotage wasn't about silencing my inner critic, but rather composing a new symphony, one where my voice resonates with authenticity, confidence, and the unwavering belief in my right to lead. The orchestra of doubt still plays, but its melody is now drowned out by the powerful crescendo of my own voice, amplified by the allies, risks, and self-belief that paved my path.

This is not just my story, but a call to action. To every leader who has ever struggled with self-doubt, to every voice yearning to be heard. Remember: You belong. Your voice matters. Own it, amplify it, lead the symphony. It won't be smooth sailing. But the rewards? Priceless.

My early skepticism had metamorphosed into a powerful tool, guiding me towards becoming a subject matter expert. Sharing these stories of my senior leadership transition, I realize that even the most "uncomfortable" learning experiences can pave the way for remarkable growth. For within the pages of forgotten encyclopedias and heated family debates lay the seeds of a leader.

Parisia Clyburn-Patton, referred by many in the field as O.P.T., the "Olivia Pope of Technology," is an Enterprise Tech Risk Governance & Controls Officer for Truist Financial Corporation, and a proud HBCU grad of South Carolina State University. From managing computer labs in the music halls of the Cincinnati Conservatory, to a Tech MBA from Georgia Tech, Parisia has been a global tech and risk leader for 25 years, driving large-scale tech transformation across multiple industries and disciplines.

Crafting Your Legacy: Building Your Brand for Long-Term Impact and Sustainability

by Ebrima Ceesay & Morgan Tucker

Morgan Tucker: In 2019, after a decade with Accenture, I was promoted to managing director. I felt I'd done more than survive; I'd thrived. Risen above. Arrived! Certainly, my excellence had finally been acknowledged and rewarded and the reputation I'd cultivated as an engagement delivery turnaround artist—with strong followership and more advocates and sponsors than any one person could expect—was well known amongst peers and leadership alike. Everyone knew who I was, why I was the right choice for promotion, and why I had to be the right cybersecurity leader to own one of Accenture's largest accounts in its North American Resources business. Right?

Guess what? A month after I was promoted, the lead for our security business across the Southern U.S. called with a proposition to have me leave our resources cybersecurity practice to come lead security for our products client portfolio in the South. New industry, new client cybersecurity challenges, new lingo and industry parlance, new boss, new success metrics, new (and heightened) scrutiny. I knew for sure, what had gotten me to my current post wasn't going to get me *"there."* To be successful in the new role, I had to think differently, had to move differently, I had to *be* a different security practitioner.

Over the previous four years, the one constant I'd been able to trade on was the brand I'd developed—a consistently excellent facilitator of intimate, vulnerable, and meaningful dialogue amongst clients and colleagues alike, a connector of dots, and "the guy who knows *everyone."* Plus, the ability to identify and address people challenges that underlie surface programmatic issues.

Jeff Bezos, the founder of Amazon, is famously quoted as saying, "Your brand is what people say about you when you're not in the room." Effective personal branding differentiates you from the pack, the easiest answer to the questions, "What should I call him/her for?" and "Why are they the best person for the job?" We must communicate clearly who we are and what we do, or risk being misunderstood, mischaracterized, and missing out on opportunity.

Personal branding for senior executives (especially in non-diverse environments), and the intersection of personal branding with racial and cultural identity.

Executive presence is probably the aspect of my brand that I've evolved with the most intent. I say "evolved" and "intent" pointedly here because I've counseled and mentored many people who seemed to think that once you've acquired a decent amount of experience, and are able to articulate your brand, and have validated that the rest of the world similarly perceives that brand—that's it. Not true. Your brand must evolve; the only constant, after all, is change.

For IT executives getting a whiff of the rarified air in the C-Suite and senior leadership ranks, it's critical to acknowledge that the roles we'll occupy next will likely require trading in some of the traits that brought us success, visibility, and recognition thus far.

Senior leaders of color need to consider how (or whether) ethnicity comes through as a component of brand. While there's no perfect way to be a Black senior executive, some of us try hard to distance ourselves from what could be a key differentiator; seeing people who look like me making this personal choice has always made me uneasy in ways that aren't easy to articulate.

But, thinking about the most senior Black executive leaders in my business—specifically, Marty Rodgers, CEO for all of Accenture's business in the Southern U.S.—I've only ever seen differentiated strength in how unflinchingly they incorporate the impact of their unique lived experiences into how they lead. As I see it, doing so with integrity and authenticity earns them the kind of credibility and followership that others (who lack the distinction of being Black and excellent) would die for.

Assessing your current brand.

If you're reading this, you've no doubt spent time thinking about and developing your personal brand. If you're a senior executive leader, you've spent years if not decades establishing, articulating, refining, recalibrating, reviewing, and refreshing that brand. But have you taken time to perform your own periodic "brand audit" like your favorite consumer brands periodically do? I'm talking about doing a third-party audit of your brand. Yes, I mean asking someone else for their thoughts on what your brand is in your organization and whether it truly resonates. (If you're already fearlessly consistent about asking for, sitting through, and incorporating feedback from your teams, your peers, and your leadership, please feel free to skip this section and go grab a cup of coffee. If, however, like me, you were a bit shocked after having the word "executive" appended to a new title, then stick around.)

It wasn't until very recently that I began to act according to the mantra I've spouted to others for so long: that feedback is a gift. If I'm honest, my imposter wouldn't allow me to proactively ask others "How am I doing?" or "What would you say are my strengths and where are my opportunities to grow?" It was my imposter whispering in my ear the same tired, negative narrative about what a fraud I was, how I had only arrived at my position after having flown under the radar for years.

Fortunately for me, I'd married a friend and consort who, well before I ever would have, grew tired of hearing my incessant self-deprecating mumbling and, in a moment of sheer exasperation, asked me, "Morgan, do you really think you're so smart that somehow you've outwitted all these people all this time and conned them into promoting you? Or perhaps it's possible that you actually are exactly who they think you are: a talented, high-performing leader with a stellar reputation and a ridiculously strong and vocal followership?"

What could I even say to that? So, I said nothing. Instead, I took the feedback from my wife, and I still use it today to push past all the reasons (excuses) for why I shouldn't ever ask anyone to tell me how they perceive my personal brand.

But I digress.

So then, now that my wife's rhetorical question has hopefully helped you see why you too should be fearless about conducting your own brand audit—how exactly do you do one?

Elements of a brand audit.

Part of the value of auditing your personal brand is squaring your perception of yourself with how the rest of the world sees you. Which means first taking stock of what *you* believe to be your differentiators and opportunities to grow. If it helps (and how could it not?), take inventory and write it all down. When I do this for myself (shooting for an annual audit, but every two years at worst), it feels a lot like a SWOT Analysis (Strengths, Weaknesses, Opportunities, Threats). Tactically, I'm asking myself questions about my:

Strengths

- Where is the empirical evidence (aka the receipts) to corroborate that I really am that guy?
- How/why do I stand out in my peer group? What are my competitive advantages?

Weaknesses

- What are the things that might be holding me back from *better*?
- What are some of the less savory things I've heard about myself from peers/friends/family/leaders either directly or indirectly?

Opportunities

- What access do I have to resources (mentors, sponsors, advocates, friends, allies, visibility, finance) that I can accelerate toward my goals and/or correct course on identified areas for growth?

Threats

(Candidly, I tend to find thinking about threats not as useful to me; your mileage may vary.)

- What competitors might box me out of the space I need to be in?
- How are they preparing themselves to enter that space? Am I equally, if not better prepared?
- What access to resources do they have that I might need, and how do I close that gap?

Once my personal inventory is completed, I'm halfway through the exercise. Now the real work begins and it's time to hear from those who matter most: senior leaders in an organization, mentors, peers (not just those you love and trust), even family members who are tuned into your executive leadership journey.

Crafting a brand vision that aligns with personal values and career aspirations.

Ebrima Ceesay: To craft a brand vision that aligns with my personal values and career aspirations, I reflect deeply on what I stand for and where I see myself in the future. I identify core values that are non-negotiable and aspirations that drive me. Then, I articulate a vision statement that encapsulates these elements, so it can serve as a guiding light for all my actions and decisions. This process ensures my brand remains authentic to who I am while strategically positioning me towards my desired career trajectory.

For someone passionate about cybersecurity and AI/ML, my brand vision is "To lead in the integration of AI/ML technologies within cybersecurity, creating smarter, adaptive security measures that protect against evolving threats. I aim to leverage my expertise in AI/ML to innovate cybersecurity practices, ensuring businesses can thrive in a secure digital environment."

Balancing authenticity with professional expectations.

Balancing authenticity with professional expectations involves careful consideration of context and audience. While staying true to my values and personality, I adapt my communication style to suit formal business environments. While I may share personal anecdotes to illustrate a point in a board meeting, I ensure these stories have a clear link to business objectives and are delivered with professionalism. Thus, I maintain my authenticity while meeting professional standards, fostering genuine connections without compromising the decorum expected.

Setting long-term goals for your brand's evolution.

To set long-term goals for my brand's evolution, I start by envisioning where I want my brand to be in the next five to ten years, considering market trends, my personal aspirations, and potential challenges. I then break down this vision into achievable objectives, such as: expanding my digital presence, engaging in continuous learning to stay ahead in my field, and building a network of like-minded professionals. I make these goals SMART (Specific, Measurable, Achievable, Relevant, Time-bound) to track my progress and adjust my strategies as needed, ensuring my brand remains dynamic and relevant.

Effective communication strategies that help establish brand.
Morgan Tucker: Effective communication strategies to establish a personal brand include being authentic and consistent in messaging, engaging with your audience through storytelling, actively listening to and responding to feedback, leveraging social media platforms strategically, and showcasing your expertise and values. Additionally, networking and building relationships are key, as is maintaining a professional image across all interactions. Here's what I focus on.

- **Authenticity**. Ensure that my communication reflects my true values and personality. In blog posts, I share personal stories that resonate with my core beliefs.
- **Consistency.** Across all platforms, from LinkedIn to X (formerly Twitter), maintain a consistent tone and message, reinforcing brand identity as a reliable source in my field.
- **Storytelling.** Engage my audience with compelling narratives about my experiences and lessons learned, making complex topics accessible and relatable.
- **Active Listening.** Show my audience that their opinions matter by asking for feedback and responding thoughtfully, fostering two-way conversation.
- **Strategic Social Media Use.** Choose platforms that align with my brand and goals; for me that's Instagram to share visual stories of my work, and X for industry insights and discussions.
- **Expertise and Values Showcasing.** Through webinars and podcasts, share knowledge on specific topics, demonstrating my expertise and commitment to continuous learning.

- **Networking.** Actively participate in industry forums and events, online and in person, to connect with like-minded professionals and expand my reach.
- **Professional Image.** Ensure that all my public profiles and communications reflect the professional image I want to project, from the quality of content to the way I interact with others online.

Navigating brand communication in diverse corporate settings.
Ebrima Ceesay: In diverse corporate settings, I adapt my brand communication by understanding the cultural and organizational context of each audience. Navigating brand communication in diverse settings involves keen observation and adaptability. If presenting to a more traditional company, I opt for a formal tone and focus on how my expertise aligns with their values. In a startup environment, I adopt a more casual and innovative approach, sharing insights on cutting-edge trends. When engaging with a multinational company, I emphasize global trends and inclusivity, with examples that span different cultures. But if interacting with a local business, I focus on community values and regional success stories. In online meetings with international partners, I consciously use clear and accessible language, mindful of non-native English speakers, keeping my message inclusive. Meanwhile, during in-person meetings with local firms, I might include more colloquial expressions and local industry references, making my communication relatable and personalized.

Tailoring my message and delivery means my personal brand resonates across different corporate landscapes, fostering meaningful connections and opportunities. This approach demonstrates my versatility and shows respect and understanding for various corporate cultures, helping me to build trust across a wide array of environments.

Digital platforms.
Morgan Tucker: Much of what you'll find online with regard to personal branding is about burnishing your online/social media

presence. But let's be clear on terms and objectives; this book is meant to share experiences that will aid, abet, and accelerate the advancement of Black IT executives to C-Suite roles. We need, primarily, to be laser focused on understanding and dominating the inside game of chess that leads to searing our personal "so what?" into the brains of the senior executive leaders whose influence will put us in our organization's corner suites.

Social media platforms offer phenomenal exposure for a well-developed personal brand but over the last decade I've developed a Missouri "show me" attitude toward how folk portray themselves on the major platforms. And I'm betting senior leaders we need to sponsor our ascent to the C-Suite are the same. Say what you want on Instagram about how you lead with authenticity and your "innovation first" approach but, in the end, that's all just pretty *marketsprechen* until you've demonstrated it to those who matter most for your career progression (hint: those are not the people clicking the "like" and "follow" buttons in your SM feeds.) If digital platforms are at best a means of providing your organization's senior executives with another datapoint that corroborates your worthiness of their sponsorship, what other strategies are worth deploying to establish and scale visibility of the personal brands we've developed?

Public speaking.

The man I credit with sponsoring my promotion to managing director had, over his decades-long career as a cybersecurity consulting executive, developed so much content to draw upon for any speaking engagement, that watching him jump on stage with only the briefest of reminders as to the audience and where the conversation needed to land, was kind of like watching Tariq "Black Thought" Trotter spit completely coherent freestyle lyrical fire, on command, live.

Like BT, my colleague has a genuine gift that I don't think he ever really acknowledges. Once, over drinks the night before a presentation, I caught him griping about so-called leaders who couldn't string together a minimally coherent on-stage cybersecurity discussion. I responded that he was a natural at public speaking and that most mere mortals are not.

I was the "mere mortal" I was thinking about. While I don't detest public speaking, when it's what I've been called to do, I've always been at my best with plenty of lead time and rehearsal. However, I have discovered my jam: moderating panels and "fireside chat" style discussions. This may be because of the team sport feel of being on stage with others, all rowing in the same direction, wanting to show up well for one another. Or the comforting feeling of knowing that, for at least some of the time in the spotlight I'll get a moment to breathe, gauge how the discussion is landing with the crowd, and recalibrate. Regardless of why I prefer discussion formats over solo speaking, I know this about myself and I'm grateful for that knowledge because it's empowered me to be infinitely more proactive.

In pursuing the path to senior executive leader, moving with the knowledge of how I can make the most and best impact—playing to my strength—is critical. Public speaking is typically a must for any business leader, and whether you love it or not, engaging crowds internal and external to your organization is an opportunity to reiterate the power of your brand. I recommend raising your hand for every such opportunity because many others will not.

Thinking back to the concept of brand auditing, specifically assessing and neutralizing potential threats, proactively asking to be the one on stage is a great way to differentiate yourself from the other folk competing for the finite resource of visibility. Since I know that I love to moderate panels, I will always raise my hand first to run the show, and while everyone else is still deciding, or figuring out how to prepare the right messages, I'm already on the way to knocking this thing out of the park. The panel format plays to some of my superpowers including listening with intent; reacting to panelists with questions that tee up opportunities for them to share insights; and facilitating consistently real and impactful conversations in which the audience sees the panel's chemistry develop in real time.

Networking.

Having a personal brand but no network through which to communicate is tantamount to having no brand at all. If your brand is the sunflower, your network is the soil, sun, water, and fertilizer it needs to flourish. When the network is robust, the sunflower blossoms, opens, and goes to seed on schedule.

But first you have to make the initial investment in cultivating your network—finding fertile ground (right audience to help you achieve your ambition); good sunlight (consistent visibility to the right sponsors, mentors, advocates); planting seeds (commit, show up, say what you will do, do what you say); attention to watering (access to opportunities like public speaking); tilling and aerating (staying connected); weeding (checking in, inventing reasons to be in touch if you must).

It's rarely the case that anyone's going to deed you a well-cultivated, fertilized garden plot in a prime location that's already producing. And even if that does happen (we occasionally do get lucky, though that may be another way of saying the network is working), you still need to water, weed, and occasionally prune.

How will you know when you've effectively communicated your brand? When you know for certain that people who know you are cascading this vision to others, without being prompted by you and without you in the room. The entire point of cultivating your brand is so that opportunity finds its way *to* you, and one of the surest signs that this is happening is when peers and senior leaders ring or ping to say "I'm reaching out on the advice of so-and-so who said you're the best person they know to do X,"—and X is nearly verbatim a core component of the brand you've built. These are the receipts.

Overcoming challenges and bias by building resilience and turning obstacles into opportunities for brand strengthening.

In 2017, when I was a new(ish) security senior manager, Jim, our global lead for resources cybersecurity asked me to step into a security architect role that needed to be backfilled at one of Accenture's largest and most important accounts. After some discussion about objectives and success metrics for the role, I admitted to Jim that I thought that there were a number of other colleagues whose experience would better set them up for success than mine—and that I'd do the work of aligning someone, so Jim didn't have to.

Jim thought about this for a moment before explaining. "That's totally fair. I appreciate your transparency. But here's the deal: I need somebody to go out there and show some leadership. This is a huge account; too big to fail. But we are failing because we've got a bunch of kids in suits with no one to lead them. I want you to team up with another senior manager I'm grooming to be one of security's next managing directors—and she's not presently a security person, so this is another part of why I need you. She's got a strong ERP background, and I need her to come grow our ERP security business in resources, but I need you to take her under your wing and make her a cyber person. Morgan, I need you to show some leadership to these folk."

One of the most senior people in the world's largest cybersecurity consulting services business just told me he specifically wanted me to go help turn things around on this too-big-to-fail account while also supporting the success of someone critical to the part of the business that he specifically owned. Of course, I said yes. That was late on a Wednesday afternoon and on Monday morning I was on the ground getting to know the client, the security program we were contracted to deliver, and the many, many Accenture people supporting the work. It seems consultants are paid to drink from fire hoses, and that's what I did.

It didn't take long to see what Jim meant about "kids in suits." It wasn't that there were no leaders on the account, simply that real leaders were few, outnumbered by managers masquerading as leaders, and that many of those LINOs (Leaders In Name Only) were, themselves, looking to be led. What I discovered was that at the root of almost every instance where our client had identified a "delivery challenge" or "technology problem," there was instead a leadership gap.

Like the abysmal optics of me sitting in a conference room with senior client leadership, hearing a car full of giggling, gossiping analysts in the background during a client working session because the Accenture person "leading" and the entire team went for a long offsite lunch, forgetting the start time for the meeting and having to dial in while driving. Or, there was the senior manager (Caucasian) who, upon finding out that he and a junior member of his team (African-American) were not only from the same part of the same state but that they also shared a surname, gleefully wondered in a room full of colleagues, "Hey, what do you think the chances are that one of my ancestors owned one of your people?" Racism, bullying, misogyny, bigotry, rampant entitlement, unchecked laziness—we had it all. At the core, it was these and other people problems that were challenging our ability to consistently deliver with excellence. And because Jim believed in my brand as a connector of and to people, and as a person who seeks first to understand and then to be understood so that relationships rooted in empathy and trust can grow, he knew that I was the person to help right this particular ship.

Over the next three years, righting the ship is what I did. Turning the tide on the most acute issues took less than a year—meeting people where they were, being vulnerable with them about the (many) times I'd had a leader identify my own opportunities for improvement, moving some members of the team along to projects where their style might be better appreciated (or at least tolerated). For a select few, I took the additional step of seeking assistance from our employee relations colleagues to find other employers altogether.

Final Words.

Challenges, in work and in life, are opportunities to develop, market, audit, and refine our personal brands. When it comes to my own brand, it's nearly always been the case that what first looked like a major problem was, in fact, a gift because that problem gave me the spotlight and the mandate to activate my brand for the world to witness, and to test the truth and impact of what I say which makes me stand out.

Louis Pasteur, the French chemist renowned for his groundbreaking discoveries in vaccination, microbiology, and pasteurization once said, "Chance favors only the prepared mind." Had I not been honest with Jim, I would probably still have found my way to success, but I believe I would have also cut myself off from the opportunity to showcase my brand as a leader of people. Advocating for what I knew to be the power of my personal brand (I can "people" with the best of them) created the opportunity for me to show that I could see past surface challenges to address the people problems that are generally core to what's gone wrong.

Ebrima N. Ceesay is Vice President of Platform Automation and Data Science at Mastercard and a cybersecurity professional with more than18 years of extensive technical expertise and management experience in diverse aspects of cybersecurity programs, information assurance, and research and development. He has led cybersecurity initiatives in collaboration with the White House, Department of Homeland Security, and various industry stakeholders.

Morgan S. Tucker leads cybersecurity for Accenture's portfolio of Life Sciences clients across the Southern U.S. He is an executive sponsor for cybersecurity recruiting and diverse talent retention programs, an alum of Morehouse College, and has been married for 25 years with three sons.

Steering the "Ships"

by Carl Daniels

Before any discussion on leadership development, it might help to know a little about my background, experience, and journey. I have had some incredible learning experiences, and along the way I have been very thankful, humbled, and sometimes bruised. I draw strength and pride from my family and my sense of determination and perseverance is largely rooted in my mother, who is a first-generation business owner of 'Passions', a beauty salon. Before she achieved this goal, I lost my dad at the age of 12 and in reflection it was largely my mom who inspired me to be a great leader, father, and provider for my family. She started as a receptionist at a high-end boutique salon where only a few black customers were patrons (i.e. doctors or lawyers). Her clients encouraged her to "go big" after seeing her potential; she was known for using a groundbreaking technique for improving their hair. Passions Salon quickly grew into Passions Enterprises Inc, a full-service salon with a staff that serviced diverse clientele with all types of hair. My mother used tough love and sacrifice to propel me to chase my goal and dream big. Looking at my daughter, I am impressed by her drive to overachieve, and I see a lot of my mother in her determination to chase her 'passions' – even without me pushing. All of this is background to understand how I contextualize the success I have had professionally; so much of what I have accomplished is because of them.

Since leaving the Richmond, Virginia area, I've handled many big global companies spanning over 3 major business industries, IT investments, Fortune 100 customers, and new business engagements, finding my place on the high potential radar as an executive director trying to break through. When mentors or sponsors made me aware of spotlight opportunities with a "you are perfect for this" introduction, it motivated me to shoot for the stars for my family.

My 25-year career to date has included IT corporate leadership for telecommunications, management consulting, global business process outsourcing, and now as part of the senior leadership team for M&A planning and integration at a major healthcare company. So maybe I can provide a helpful perspective; you never know, even an imperfect story or example can fill in some blanks or provide some motivation, so let's go!

The relationships—the "ships"—I've like to discuss are about the kind of leader who is the first to congratulate you after a long well-deserved promotion. Or the person who gives you your first promotion to senior manager or above. Either of these might be a mentor, a sponsor, or a partner. What is a business partner? How far can a mentor take you? Is a sponsor super critical for breaking through ceilings (glass or otherwise)?

Mentorships.

A mentor is your professional coach and often serves as a trusted counselor to help to guide and direct you to success in your professional career. When I was an intern, my first manager became my first mentor. She hand-selected for me—the only Black summer intern. I will never forget her face when I met her at 7:59 on day one. She was a highly respected manager, soon to be district manager, a first-generation college graduate, and leader in her family, an overachieving Black female. Let's call her Suzanne. This upwardly mobile Manager was professionally tough but fair and created a healthy level of fear and respect in her organization. She was extremely structured and precisely followed policy, procedures, and guidelines; a perfect fit for me because you know, kids need structure. I came in a little early, stayed late, and completed my technical requirement and summaries with high marks from Suzanne's team. This included a summary of the file retrieval and print procedures for a new Apple PC that was just sitting in a corner with limited use.

I think Suzanne was extremely surprised that her summer intern was a Black male and was proud of me by summer's end. Her handshake and hug were priceless. That experience had me thinking this was all that I needed to succeed. The experience of completing my summer internship and getting a job offer around the same time as the top students in our business school was enough to drop me to my knees with thanks from above. When I finally came on board as a full-time developer, I thought I had the playbook for continued career growth. I excelled with two very competitive promotions to senior specialist by year five, and to manager by year eight. The district manager kept an eye on me as I moved to different areas within this Fortune 100 company.

If you have a potential mentor—formal or informal—or one you've targeted, get them on your reoccurring calendar, about every other month or once each quarter on average. Consider adding a new (likely informal) mentor after any major organization change. I recommend that you have at least one mentor in your department to help navigate the energy, fire, culture, and confidence we all have.

Your mentor can act like a life coach, especially early in a career. Effective mentors care about you as a person, not only your professional life. Try to get as much direct and honest feedback on your leadership potential, style, and presence for the next level. A mentor may ask often about your family with sincerity. A mentor should know your tendencies, pros and cons, and identify any needed career development—someone curious about you and your potential, with great ideas for your future. You should feel comfortable asking them almost anything. An assigned mentor relationship may be less personal, but you still want to feel comfortable asking about the good, the bad, and the messy political at work.

Be sure to thank your mentors for their extra time and consider sending them a heartfelt hand-written thank card in at least the first year.

Sponsorships.

A sponsor can also be a mentor, but there's a distinction. At the highest levels of executive leadership, you are normally pulled up or referred without even applying for jobs. A sponsor is the person who makes things happen at the higher table to "pull you up," and if you fail, that reflects on the sponsor.

Some leadership theories are heavily weighted on finding a sponsor once at or near an executive level, because by then maybe we should know how to play the game without heavy mentoring. Two good sources on this topic are *Forget a Mentor, Find a Sponsor: The New Way to Fast-Track your Career* by Sylvia Ann Hewlett and *The Leadership Pipeline: How to Build the Leadership-Powered Company* by Ram Charan.

My first sponsor who pulled me into the management ranks was Rose; it was a bold move though I believe everyone knew it was just a matter of time. Despite several successful customer implementations, I was not perfect, but I was ready. She had a way of looking at me when I made a mistake that still brings a little fear. I had to find the balance of high risk and high reward while still being one of the best all-around professionals in our IT department. I felt the weight of people counting on me to succeed and I never wanted to disappoint Rose. Now, I want anyone who I sponsor to feel the same way, to be strong enough to persevere and make me proud.

Being a good soldier and top professional who is reliable, trusted, consistent and faithful is so important. Being open to developmental or improvement feedback instead of positive feedback is also crucial. A sponsor must know their feedback will translate into quick results that reduce risk. The ratio of *good* to *get better* feedback from a new sponsor could be 30/70. They need and expect you to shine under the bright lights or when deep in the pool, to live up to the hype about you without high maintenance. You will flip that ratio of feedback when you make them look good, and chances are you and that sponsor will follow each other throughout your careers to even bigger things.

A sponsor may one day be the former colleague who pulls you into a C-Suite position, or as a colleague at a new firm—a major *external sponsor*. This pattern is well known when new executives join a firm, bringing along trusted cronies. This is a *get you in the door sponsor*. As a leader you will one day could do this, helping "our people." Occasionally, someone will let you down, but don't stop helping our people because of one bad actor. Stay inspired, try not to lose your hope, and place better bets.

A sponsor is a powerful blessing that may come along when you least expect. Important people are watching you and how you handle your formal and informal professional *and* personal relationships. Stay present, insightful, and be relevant in large forums where your expertise is expected but don't force it; don't lose confidence, focus, or insight after tough projects or revenue quarters. Instead, offer solutions or leadership insight where needed; these qualities increase your chance of being "sponsorable." A good sponsor will see you on their dream team or someone else's and wants to see you get your due; they like investing in you as a person. Be sure to thank them for their extra time.

Partnerships.

Finding good business partners who can enhance your ability, opportunities, and potential is also important. I have had a variety of useful partnerships throughout my career, including folks who were a part of my travel team for top client meetings, to close business or project team members, random leaders or managers within the company, and business process owners. Because I was comfortable speaking with all types of folks from a variety of backgrounds, I never felt uncomfortable with anyone in a business climate. Instead, I was especially curious about the extremely confident, eccentric, and maybe even intimidating folks who others sometimes avoid. Perhaps in turn, there was a certain curiosity about me (as in, how did *he* get here?) and why I seemed so calm or confident most of the time. I was just being myself, keeping my eyes forward with seemingly few distractions. As I gained traction in my career, I expanded my circle beyond only my trusted partners. If your eyes said you were a nice person, I might give a smile or nod while keeping it moving to the next meeting. Good partners can include nice and maybe not-so-nice people, as long as they can bring value to the revenue table while being reasonably trustworthy, consistent, and reliable.

It has helped me to stay approachable and easy to engage for important business Q&A especially after formal readouts. I've received priceless nuggets of information throughout my career, unsolicited, during casual conversation *just because I was there*. I learned more and more about the game of upward mobility, strategic client relationships, and self-marketing this way each year.

There are partners who may not like you or trust you either, by the way. They can give you the best unsolicited feedback on the spot followed by "but I really respect you." I worked together with one such person to give the best impromptu presentations for the toughest clients and we grew to admire each other.

I highly recommend Harvey J. Coleman's book *Empowering Yourself: The Organizational Game Revealed* as a foundational playbook on unwritten rules of relationships. While the foundational structure information on our "people game" is timeless, and even though you may find some references dated, you can still apply them today through the "gameboard" he outlines. His book is like Ben Hogan's *Five Lessons: The Modern Fundamentals of Golf* published back in 1957, yet still one of the bibles for basic golf swing, easily adjusted for today's equipment.

Consider your partners carefully; a match is not a matter of common values, personalities, race, or religious background. A good partner is generally an important player in a high stakes game who values what you might bring to the table and vice versa. Partnerships are generally two-way streets, but also consider being a great partner without expecting anything in return, an always-appreciated gesture (well, most of the time). What's most important is forming partnerships with strategic goals defined, attaining envious achievements, and personal growth without wavering. That is when "ships" sail smoothly in all types of waters. These "ships" may carry us to big new opportunities we had not envisioned.

Carl Daniels brings over 25 years of IT leadership at top industry leaders for telecommunications, management consulting, business process outsourcing, marketing, and now healthcare services. He has successfully led portfolio project management, application development, infrastructure, operations, and M&A integration for business partnership with technology. Carl enjoys golf and outdoor activities along with his daughter in the southwest region on weekends.

Learning the Value of Support Systems and Networks on the Corporate Ascent

by Judy Covington-Poole

In building my technology career, I have always aspired for financial gain to support my family. I grew up in a home where my mother was the "side chick" and learned at a very young age the importance of knowing and demonstrating my value. In school I pushed myself to be the best, graduating third in my class from William Marion Raines High School in Jacksonville, Florida. Today I look back and think, I should have worked harder. Professionally, I approached work with that same tenacity—often the first person in the office and last to leave, pushing to be recognized as the best at what I did.

Only much later in my career did I see the value of having a support system. No matter how excellent my performance, I still needed a cheering squad. To be successful, I needed to compete but not compare. I focused not only on my development but also the development of others around me, learning the importance of building relationships. Instead of only financial gain and money, I began to place equal emphasis on creating and maintaining authentic and healthy relationships, obtaining positions in an authentic way. But first, for me to reach a point of love, truth, and prosperity, I had to overcome my fears.

I grew up in the seventies, in an environment where African Americans dreamed but the dreams were deferred. I remember the first local Black owned skating rink, Skate City, where my friends and I all went on Fridays and Saturdays. We had a blast. Then this business, like all the businesses in our community, failed. I must have taken it all in and later on, recognized my motivations were driven by fear of failure. I should have been driven by the faith of success.

I started hearing my own voice when I left my first corporate position at Florida Blue Cross and Blue Shield. I stepped out on faith and a nice compensation package but was able to land my first consulting position with CitiCard. Yet I did not know my worth. I was being underpaid. It was time to start building relationships with other consultants. They taught me the following: "You have a pimp and make sure your pimp is paying you well." With this knowledge, I began to learn the art of negotiation and landed consulting fees three times what I was originally making. It was not just about the work I was delivering; it was knowing the value of me being in the room at those corporations.

This is an example of how I started building my technology network though I did not yet know my path would lead me to other opportunities. I was focused on my deliveries and my earnings. When I accepted a position at Fidelity National Financial, I did not know about the conversations people had behind closed doors. Doing the right things in a technical space was what drove me, and my clients and customers respected me. However, I was burning bridges with my leadership. How? I was not visible enough. I was not "kissing babies and shaking hands." I was unaware that clients provided bonuses to other consultants, apparently in secret, to those who built relationships with them and invested the time. I had to adjust my paradigm. What do I want out of corporate America? What am I willing to sacrifice?

I could not answer the questions until I was clear on my goals. By then my goals by then included wanting to be a leader who leads with integrity and passion. That goal became reality when I started focusing on helping others, assisting with resumes, mentorship, and providing references. This was challenging for me. Why? Once, I reached out to an African American Vice President in the organization about a promotional opportunity in his area. He did not respond but sent one of his "people" to talk to me. That really set me back in trusting others. Why do we have to be so guarded with helping other people of color?

African American woman in corporate America face judgment and other factors that derail us from aspiring to grow our careers. If a white male had taken the same approach, he likely would have been rewarded for his ambition. These are the traumas we learn to deal with, and we should not. This is when I realized I had to focus on relationships of authenticity, and it helped me to learn the type of leader I did *not* want to be.

~~~

Take stock of your existing contacts, both online and offline. This includes friends, family, colleagues, classmates, mentors, and acquaintances. Evaluate the strength of your connections and their relevance to your goals. I do not believe in dismissing human beings; however, there is a season for every relationship. My motto is to take the best others have to offer and leave the garbage with them. I had to do this in my personal life and my professional life. In my personal life, even with family, when I am visioning taking on a new assignment, at first, I go quiet about it and do not confide in anyone because when my career started to elevate, I noticed certain friends playing the one-upmanship game (perhaps not out of jealousy, but I needed more growth and maturity).

Today, my network is always evolving; my mindset needs to be aligned with people of a similar like-mindedness related to generosity and thankfulness. Being around individuals who consistently take, speak only about themselves, and otherwise exhibit selfishness will not take me where I aspire to be. I aim to build a diverse network that includes people from various industries, backgrounds, and perspectives, which enriches and provides a broad range of insights and opportunities. If you have a mix of young, old, Black, white, Asian, African, Caribbean people around, you get a diverse dialogue to help challenge the paradigms you have learned in your life since childhood. My best friend today is a man from a small village in India. He and I worked together several years ago. When he invited me to India to visit him and his family, I was surprised when I said yes and booked my ticket. We are not just business partners, we are family.

Networking is not just about asking for favors; it's about building genuine relationships. The established relationships I have today came from me being vulnerable to ask for help. At one point I was watching others progressing in their careers while I was not. I realized I needed a mentor and asked one of the leaders in the organization to be my mentor. This person gave me insights into my blind spots, counseled that I needed to take on high visibility projects and vocalize my accomplishments so others would be aware of my
~~~

growth and the value I bring. Today I have three mentors who provide insight and help me continue to grow personally and professionally.

I also learned the importance of regularly nurturing relationships by staying in touch with contacts. This could involve sending occasional emails, meeting for coffee, attending industry events together, or connecting on social media platforms. For me, this one is hard. Life takes us down roads of turns, circles, and highways. You get disjointed from others. But I have learned to be intentional with maintaining relationships through the years, making sure I feed the relationships to keep them healthy and satisfying.

To get started, identify individuals who can serve as mentors or advisors in specific areas of your life or career. Look for people who have achieved success in fields you're interested in and approach them respectfully with a clear request for guidance. Others have done it before you. Others have taken the hill and took the shots. Your mentor can help you see your blind spots. One of my mentors, who is a white male, has helped me navigate up the leadership chain, teaching me the art of researching important people in the organization and utilizing this information to establish relationships. My weakness is managing up that leadership chain which is why I engaged this person. With his leadership and guidance, I have established personal relationships with vice presidents and the CIO in our organization. This has propelled me to meetings with other VPs and CIOs at other corporations. Eventually, I realized the mentorship relationship can lead to the sponsorship relationship and I began to identify many mentors for my career and my life. I am still working to establish a sponsorship relationship. I am listening and recognizing the players and the people in the room. This means I am listening more than I am talking.

All the while, one must cultivate an online presence through platforms like LinkedIn, X (formerly Twitter), and professional forums. Share your insights, engage in discussions, and connect with like-minded individuals in your field. I am not the best at building an online presence but in today's world, this needs to be a priority. Simply having an online presence is not sufficient; you need to cultivate that presence. This is an area where I am still working to grow.

Attending industry conferences, workshops, seminars, and networking events provides valuable opportunities to meet new people, exchange ideas, and stay updated. As an introvert (and because “life be life-ing"), this is a challenge, so I place one event on my calendar each month so that I am sure to attend. This is where a support system can assist. Make plans to attend with a friend or a colleague. Once there, it’s a chance to talk with individuals you would not have ordinarily approached. Take the risk and make a connection.

As you progress in your career, look for ways to give back to your network. Offer mentorship to others, make introductions, share opportunities. Building a reputation as a generous and supportive member of your network will enhance your relationships and open doors in the future. I believe that the universe will position you where you need to be, at the time you need to be there. But first you have to be prepared for the overflow and ready for the opportunity. The energy you put out is the energy you will receive back.

~~~

According to personality tests, my personality type is an INTJ, defined as a Quiet Specialist –an individual who is a logical thinker, observing and exploring how the world works. The behavioral characteristics for this personality type are: a private individual who is intense and cerebral; perceptive, innovative, secretive, and isolated. Many people might look at that description and say, “That is not you!” But it *is* me. It is not natural for me to speak up or cultivate relationships. Those are skills I have worked hard to develop, but they are not intuitive or comfortable. I have struggled; but through that struggle I have learned to use my superpowers.

I became the executive administrator for our CIO, as well as director of product engineering. In the role as EA, I had to lead the technology organization holistically. I developed a relationship with two CIOs, the previous one and the new one. In this role, I had to speak up and share the state of our development teams and communicate holistically on goals. My CIO once said, “Judy, you are a natural leader. Do not forget who you are.” This helped me
~~~

recognize I can utilize my natural style of leadership and enhance my skills to be a more effective leader at any organization, supporting any industry.

I had to identify my superpowers; metaphorically, I learned to put on my pearls, step into my heels, and have interactive conversations with people at all levels. Building relationships in places that I would have never conceived. Realizing I had a voice and an opinion worth sharing broke through my hesitation and nervousness. This, coupled with my Southern characteristics—being humble, respectful—freed me to utilize my interests to gain the interests of others. For example, when I would meet with male leaders, I found I could break barriers by discussing SEC college football and teams' on-field strategies, which provided others with some insight into my own strategic and tactical thinking. Or I would share my faith and my spirituality, allowing other leaders to recognize my positivity and creativity.

I remain walking in my truth: that I am no better than anyone else, however, no one is better than me because I am wonderfully and uniquely made. God has enriched me and loves me in the same capacity in which He loves us all.

Judy Covington-Poole is a Director at CSX in the technology division. She is a leader in the technology industry with over 25 years of IT experience in insurance, healthcare, banking, retail, and logistics. Judy is a dynamic executive who successfully develops and leads teams.

My Transition to Senior Management: Black Man's Perspective

by Perry Lee Woodard

Transitioning from an individual contributor to a senior management role is a challenging and complex journey for the best of us, filled with trials and tribulations. However, for a Black male, this transition can be even more complex and challenging, as we are faced with a unique set of barriers and obstacles.

For me as a Black male, the transition was marked by a significant shift in responsibilities, expectations, and leadership dynamics. Initially I struggled to find my footing. I had been an excellent individual contributor where my focus was primarily on executing tasks, meeting deadlines, and excelling in a specific area of expertise. However, the move into senior management requires a broader skill set including strategic thinking, decision-making, team leadership, and the ability to drive organizational change.

This transition is often compounded by systemic challenges stemming from historical barriers that can manifest in various forms: limited access to professional networks; unconscious bias in the workplace; lack of mentorship opportunities; unequal access to career advancement resources. As a result, I faced a steeper uphill battle as I strove to reach senior management positions. If it is a white-male dominated space, then the chosen ones are typically white males. Nevertheless, we must find a way to overcome and expose our greatness.

One of the biggest and most serious errors that I made while pursuing senior leadership roles was to assume that my competence and qualifications were enough to get me noticed. Despite possessing the required skills and experience, by my account, I found myself scrutinized more rigorously, subjected to unfair performance evaluations, or overlooked for promotions and opportunities. At the time, I attributed these to the broader issues of systemic bias and racial inequality within the corporate landscape. Even so, these small setbacks and slow advancements had a significant impact on my career trajectory.

But first, let's back up. As an early IT Professional, I had remarkable career success, with accomplishments that were highly technical. I created and led major complex infrastructure designs and implementations. During one such program, I was asked to provide an architectural overview and business impact of the solution. This white paper was being delivered to the company's CIO. I found immensely proud while writing and submitting this paper to my (white male) director. About a week later, the water-cooler talk was around how successful my director was and how well *his* white paper was received. Yes, his white paper! Turned out, he made one edit to my paper, removing my name and title and replacing it with his own. Could this have been a case of a shifty boss who might do this to *any* junior teammate? Perhaps. But it didn't feel that way.

As I progressed in my career, the absence of a diverse leadership model and lack of role models hindered my creativity, development, and advancement. Furthermore, the isolation and lack of a supportive network contributed to feelings of alienation and disillusionment, making the journey to senior management a tough one. Despite all of this, we do find a way to put our best foot forward. We highlight our strengths instead of lowering ourselves to our weaknesses. We focus on winning instead of merely recognizing challenges.

In another instance, while playing the role of team leader at the largest telephone company in the Unites States, the sand shifted from beneath my feet. Had I over-stepped? My senior leader (white male) made an organizational update and in the end, all minorities reporting to him (brown, Black, yellow) were transitioned to the quality assurance team under my leadership. The more respectful, promotable roles were given to young, junior IT professionals, all white males and white females.

In response, I quickly scheduled a one-on-one with my senior leader the very same day. I introduced the idea that many will see his organizational update as racist in intent and suggested that he tweak the organizational chart by introducing diversity within the groups. He said he respected my thoughts and would review. Two days later, he announced that Tom (22-year-old white male, one year out of University of Texas) was the new team lead of the QA team. Yes, he'd replaced me with a junior IT professional. Why? Well, I suppose simply because he could. At the time, I wanted to prove that I would shine and be well accomplished in any role, so I took the hit. Meanwhile, I worked at ensuring the growth of my own network.

Is there a pattern? Or just excuses on my part? The stereotypes associated with Black males in the workplace do create hurdles. Navigating the fine line between assertiveness and aggression, confidence and arrogance, or leadership and intimidation was a delicate balancing act for me as I aimed for senior management roles. The burden of disproving negative stereotypes and misconceptions added another layer of complexity to my professional journey.

The hurdles placed in front of me were not exclusively put there by white males. Several white females also attempted to burden me with roadblocks. At one point, as a new manager, it happened that all my peers and my direct leader were white females. This felt a little strange at first, but I thought I would quickly adapt. In group planning sessions, I often found myself coming up with solutions very quickly. But, as I tried to walk my peers through what I thought was obvious, my leader would interrupt me to "allow others to speak." This went on and on. In the end, often we came back to my original idea. It was so very frustrating for me. While this may be yet another instance of a toxic boss with terrible collaboration skills who just happened to be operating in an unbalanced gender and racial environment, it felt more personal to me. To be disrespected as the only Black male and only male in the team caused me to seek other groups to contribute my creativity, and I quickly moved to a different team I felt would have space for my thoughts and opinion. This is not always available as a solution.

Eventually, the lack of equitable opportunities created a need for the pendulum to swing hard in the opposite direction. Many companies have created DEI efforts and Chief Diversity Officers now monitor the lack of opportunity. Despite these efforts, Black males are often still faced with systemic barriers hindering our progress. If I am underrepresented in leadership positions, it creates a gap that limits my access to decision making policies. Over my career, I have felt isolated, and without mentorship and better sponsorship from leaders who look like me, my career opportunities were limiting.

What truly upsets me is the smooth transition into microaggressions by white males. These are the white male leaders that portray themselves as a friend or mentor yet at every opportunity are more than willing to slip in a joke at my expense. Such so-called "little humor" creates a faint concern in the minds of senior leadership about my capabilities and dedication. Subtle comments and harmful dismissive words are being utilized to indicate to all that I am not ready.

Some examples include when white people repronounce my words *minus* my Southern accent; make a comment that they could have done more with my assignment; note that my body-language stated something negative when I was actually in deep-design thinking mode. Yes, these things happened all the time during my career. Thus, at every opportunity, I would stop the meeting, and ask for clarification of such comments and for that person to validate their intent, and also explain how the comment was received by me. Afterwards, I would reach out to my corporate human resources department or representative to level-set what I heard and how it made me feel in the room.

These microaggressions create a culture of lack of respect and inequity. Whereas I am not a firm believer of numbers-based DEI programs, I am a strong advocate of mentorship and sponsorship programs targeting young, talented Black men with high potential. Thus, DEI has its place in corporate America. By acknowledging and addressing the kind of systemic barriers and biases that I faced as a Black male in the IT Professional world, companies can create more equitable environments.

How many times have I been the second choice in the selection process for promotions? I often wondered, what is the one thing that causes me to be the runner-up for the big promotion or that break-through new job? In the end, I get the "You are great but…"

speech. No negative feedback, but also, no job. So very often, when white males do not have a reason for not selecting me, the inside word is they went with their gut. Their gut nearly always turned out to be a white male with remarkably similar traits—same background, similar colleges, same self-image. I get it, I think(?). Most executives (who are mostly white males) look for themselves in candidates. While some are coming to the realization that diversity is especially important in selecting direct reports, it's a case of too little, too late for me.

In one role, as a senior manager, throughout the year I was praised by my peers for leadership and efficient improvements. Weekly, I heard words of acknowledgement on how advanced and approachable my team was since I started as their leader. During this year, executive leadership changed three times. From January through August, I reported to my executive sponsor (VP) who spoke of how quickly he was planning to promote me and had gained the support of his peers to do so. In August, he passed away from a heart attack while on vacation. This man was a friend as well as my mentor and sponsor. By default, for the next three weeks, I reported to the CIO (Indian male). Shortly thereafter, I moved to the Transformation Leader (Managing Director) – white male), reporting to him for the last three months of the year.

During this time, my company performance cycle was two months of reviews and salary adjustments, effective in the third month. As time passed with no performance review meeting scheduled, I did not think much about it. Every comment that year had been praising my efforts and accomplishments.

In late March, with no adjustment in my pay stub, I logged online to find I had received a 1% raise. At this company, 1% or 0% is utilized when a "performance improvement plan" (PIP) is required. A quick discussion with HR revealed that the CIO himself (to whom I reported for three weeks), had reduced my adjustment due to 1) lack of evidence of being a strategic thinker and 2) being reactive not proactive.

These comments were in direct contrast to all written and verbal feedback during the year. No written PIP and not even a year-end performance assessment had been completed. I could only assume he changed the percentage because he could. My transformation leader's explanation? I fell through the cracks during my move to finance. Later, I'd investigate and learn that most (not all) Black leaders received significantly lower salary increases under our new leader. Pattern?

Black males face numerous challenges like these, and more. Despite these obstacles, I (we) must break barriers and shatter stereotypes, as we make significant strides in our journey to senior management. By doing so, we are recognizing and addressing the systemic barriers that impede our progress and open doors for others.

Perry Lee Woodard has 40 years of experience and is the Senior Director of Supplier Management & Procurement at American Airlines. In this role, he is responsible for all technology spend categories and contracts (licenses, hardware, software, SaaS, and labor).

A Journey of Aha Moments on the Climb Towards Executive Leadership

by Lorenzo Keith

We tend to think of aha moments as one-time, almost instantaneous events that present an opportunity for us to reflect and learn another one of life's little lessons. And many are. But just as often, the "aha" is cumulative, spread over many moments, or an entire situation or experience. Either can impart little nuggets with profound meaning that can go on to shape who we become and provide critical fuel for life's journey. You never know when an aha! Will pop up in your professional life, but after the experience you are forever gifted and changed. If you're lucky, as long as you live you will encounter aha light bulbs; while yours may be different from mine, when you do happen for you, I encourage you to freely share them with others. Considering the road we travel to success will have many sharp turns and potholes, you never know who might need to hear your story.

Aha: A chance encounter – Merrill and Carmela.
Lesson learned: *People are people, irrespective of race, gender, socioeconomic status, or any other subjective perception, and all provide opportunities for teachable moments.*

During the 1980s, when I started my professional career, I found myself in a somewhat strange world. I had well-established relationships with family and friends, and foundational experiences gained from my educational paths. All these relationships helped me to develop confidence in myself and taught me how to socialize with others and most importantly, how to gain understanding and acquire knowledge. But little of that fully prepared me for the social and professional demands of corporate America, an environment where most everyone was more interested in my technical expertise and ability to positively impact the bottom line, than in who I was as a person. Any effort to support and nurture me, or to help me fit in and thrive, was a distant secondary goal (at best). But then, as luck would have it, my first real boss, Merrill, and his chief of staff, Carmela, were heaven sent.

Having spent most of his career working for one of the big six accounting firms, Merrill was a true professional and willing mentor. His wardrobe looked to be from Brooks Brothers, like most high-level male consultants of that day. I was coming from the opposite end of the clothing spectrum. Clearly, I had to step up my fashion game. So as my budget would allow, I promptly invested in a fashionable four-piece suit. Yes, you read that correctly, a *four*-piece suit (gray jacket, gray *and* black slacks, and a reversible gray and black vest) that provided me with enough options to cover every day of the working week plus Saturday and Sunday—a fashionista for sure! I learned that you *can* fit in, even if your fashion sense and budget are somewhat compromised.

Merrill was also a strictly by-the-books kind of guy, and early on, it seemed to me he had little to no outward personality. Carmela, his chief of staff, a former schoolteacher from rural Kentucky, immediately took a liking to me; I guess she found our social differences at that time interesting, with me being a young Black man. Both Merrill and Carmela were Caucasian, and we came from very different worlds, but it was precisely our differences that made for some of the best lessons. I can't begin to tell you how much I learned from them.

In addition to some fashion tips, I also learned some much needed dos and don'ts for professional settings, the norms of that era, very early in my professional career. Merrill emphasized meeting preparations, agendas, and managing outcomes, where Carmela was great at writing and presenting a story that maintained audience attention while supporting creative thinking. They both supported me and helped me to develop a foundation that my college life and working-class family had been ill prepared to provide. Our daily conversations getting to know them opened me up to a new world. The initial mask that I thought I saw them wearing slowly peeled away as we gained trust and learned from each other. Both became lifelong friends.

The message here is that you can and will learn from all kinds of relationships. Good or bad, learning what to do, what not to do, how to conduct yourself, plays a major part in who you become. Put all your life lessons in your basket; trust me, you'll need them later. From Merrill and Carmela, I gained true confidence in professional settings and an appreciation for embracing who I am. I look back on that experience and like to think it was DE&I before that was even a thing.

Aha: Find yourself sponsors and mentors along the way.
Lesson learned: *It pays to be liked and have much needed support and guidance, in both your social and professional journey.*

Most of us were raised to believe that the level we reach in life is determined by preparation, our character, and ability to contribute. And yes, that is true—to a point. But the world we live and work in is ever more competitive and just being good is no longer good enough. And please believe, competition only increases as you move up the ladder. Your co-workers (aka your competition) are also gifted and likely highly motivated. Like most things in life, survival of the fittest teaches us to look for the little differentiators that help us stand out. Just to be clear, without solid skills and abilities you are destined to remain at your current level or to be replaced, but those who reach the highest levels are smart enough to get an added boost from strategic supporters who help promote and provide vital advice.

That extra effort usually comes by way of someone who takes an interest in you and chooses to invest in your growth and provide strategic guidance—sponsors and mentors. (How to find sponsors and mentors is a lesson covered elsewhere in this book.) Like many good relationships, it's wise to be intentional and strategic about how we set up ourselves to attract and nurture those relationships. Target who can support your mission, and when opportunity presents itself, engage them and embrace all that the relationship has to offer. They will do for you that which you cannot do for yourself. Their knowledge and position of influence can guide you and promote you in ways that only someone who takes a liking to you will do.

I recall a time in my career when I had just reached the director level and was going through the difficulties of a significant reorganization, a real *Game of Thrones* environment. A senior leader pulled me aside and told me to find someone I could trust and respect for their strategic and political skills—valuable advice I put into action.

Once I found someone I could trust and respect, I was relieved to learn that a lot of the turmoil I was experiencing was part of a normal cycle and I had to adopt the right attitude to be a player at that table. Sponsors and mentors are always great but especially important when going through tumultuous times. Their insight, wisdom, and ability to provide that much needed helping hand is often the difference between success and failure.

Aha: Trouble can be transportation, a vehicle for personal growth.
Lesson learned: *Challenges to your personal and professional life are a certainty: faith and perseverance are the keys.*

Adversity often presents opportunities to develop resilience, creativity, and can ultimately propel you forward in your personal and professional journey. I recall a recent story told to my church congregation. In the late nineties, my pastor had just graduated from a leading divinity college in Boston at the top of his class and became aware that his home church was in the market for a pastor. Clearly this would be a dream come true if he could lead the church that he and his family had been a part of for all of his life. Beaming with pride, he applied and was promptly scheduled for a Sunday to demonstrate his abilities. Following the rousing acceptance of his sermon, one of the church elders recommended him for the position. To his surprise, one of the deacons objected to putting the direction and leadership of a 500-member congregation in the hands of such a young man. Understandably, the young minister was devastated.

Life went on. He secured a position at another church in another city, and built such an excellent reputation, he was heavily recruited by many churches, nationally. He never got the job in his home church, but he still went on to be successful; his current church just exceeded 12,000 members.

So often in my career I've felt at the end of my rope, going through a situation with a bad boss or an organization downsizing. One event came at the end of my first professional company tenure. My role was pulled out from under me by a new manager who clearly wanted someone else for its leader. I was devastated, as if my world was ending, to find myself without a job after fifteen years. So much of who I was, was tied to that organization and I began to question my abilities and self-worth. After a few months of prayer, reflection, and job search I landed a new role. And not just any new role but one that was right for the new me, setting me on a new and better path. I now see that job loss as a true blessing sent from above.

The moral of the story is, don't let challenges put in your path prevent you from achieving what you are prepared for and capable of being. Don't allow any person or organization's "No" deter you from your destiny. Just thank them for the gift of the challenge, and the fuel it provides. Learn what you can from the experience but never give up on your dreams. I'm here to tell you that most of the positive changes I have experienced in my life started from a No. A thousand no's can easily be defeated by just one yes!

Aha: Perception is often more powerful than reality.
Lesson learned: *If a tree falls in the forest and no one is there to hear, know when to make a noise.*

One reason we spend so much effort grooming ourselves is to help ensure that first impressions are positive ones—creating positive expectations, which lead to positive outcomes. As the old expression goes, if you look for positives, you often find them. And conversely, when you look for less than, you often find that. I am an introvert by nature so self-promoting is not natural for me, but in today's professional world, one must develop and look for opportunities to self-promote.

I recall a very successful project I led that resulted in significant cost savings for the company. The project was completed with limited fanfare outside my immediate organization. I was initially pleased by recognition from my leadership but later surprised that few decision makers outside of that area fully understood who was behind that success. I was further disheartened to learn that my manager was all too happy to take credit for my work during end-of-year reviews and recognitions. Another hard lesson learned.

Beat your own drum to make sure everyone can know and celebrate your accomplishments with you. That way, the drumbeat is playing in the back of their minds, reminding them of your individual recognitions and advancement. That tree falling in the forest did make a noise but maybe only a few tree squirrels heard it. If you want to rise higher, learn to make good noise and be sure others hear the noise too.

Lorenzo Keith is a technology leader with more than 30 years of diverse information technology experience developing strategic, tactical, and cost-effective solutions; leveraging human, hardware, software, and capital resources to support business objectives and customer expectations. In his current role as IT Director at Capital One, Lorenzo leads a team of IT infrastructure project managers responsible for domestic US and international portfolio deliveries.

Keys to Building Lasting Relationships

by Dom Green

We've all heard the saying that time flies when you're having fun. As I get deeper into my career (and life) I've found this to be true everywhere when you have good relationships. In fact, I'd say relationships are all that matter. In my world of managing relationships with colleagues, clients, customers, family, and friends, I've realized that life is nothing more than a culmination of key relationships. Some relationships are as finite in time as the projects or contracts that bind us together. Other relationships are laborious with a diminishing risk-reward. But the lasting relationships that span the test of time, that provide the underpinnings of a life well-lived—that make it fun—are usually joined by a few pillars. The lasting relationships in my life and career all stand upon six pillars that I want to share with you in hopes that they'll become your pillars.

Mark those achievements.
Relationships that include celebrating achievements allow us to build a strong foundation. Recognizing and commemorating each other's successes and milestones creates a sense of support and encouragement within relationships. Relationships built on sand bypass the successes of friends, family, and colleagues. Long-lasting relationships built on stone mark our mutual milestones. There is no such thing as having too many achievements to recognize.

About four months after college, I was off to the races, starting my career, when 9/11 occurred, and catapulted me into helping stand up a team of resources that would become known as TSA, the Transportation Security Administration, an organization born out of necessity under the Homeland Security Act (later the Department of Homeland Security). My role was to help manage escalations needed to clear the pathway for staffing up an agency that would provide our country with the personnel it needed to thwart another 9/11 style attack. Mine was a hyper-political, hyper-focused and demanding role with potentially dire consequences if not executed well. Calls to Congress members, governors, and our company CEO were commonplace for initiating escalations; there was no safety in quieting down the process of staffing TSA.

In my role I observed from my leaders the importance of celebrating the smallest of achievements. Whenever we onboarded one of thousands of new resources, closed an escalation with a congressional office, or simply made it another day without an incident being opened up due to a poor customer experience—we celebrated. Under the pressure of protecting national security, we still found a way to appreciate the small wins in the grand scheme. Doing so provided a pressure valve to relieve overwhelmed team members.

Since that experience, I learned to highlight and celebrate the achievements of others and their contributions; in fact, I saw that one key to building strong executive leadership relationships is to celebrate and recognize the sacrifices of others. I've thrived in corporate America for nearly 30 years now and still, a key to my success is being able to quickly establish relationships where we can all win.

When I share in others' joy and accomplishments via relationships where I've invested time and energy, it snowballs and brings my colleagues, friends, family, and clients closer together. The snowballs grow with each turn and so does a sense of unity and teamwork.

Matching up values and goals.

I am sometimes amazed at how my decision to pursue a career in engineering started me on a journey that would span multiple states and countries. Even more astonishing has been the necessity for me to recognize the values and goals of so many who didn't come from my small city on the peninsula in Virginia. Shared values and goals are crucial for harmonious and fulfilling relationships. When partners align on fundamental values and long-term objectives, it creates a strong sense of purpose and direction.

Let me put it this way. In a world full of people with varying values and goals, there were occasions when I just didn't think it would be possible to find common ground. After all I am who I want to be even if we could peel off race, gender, faith, motivations, goals, or values. I am committed to being uniquely me and I once thought that there would be too much friction involved in the search for commonality with those so dissimilar to me. So early on in my career I abandoned the search. But then I learned, that's just it. The secret sauce for any corporation is to get to that point of sharing individual values in an equitable way where no one feels left out.

I'm fortunate I picked a career path where the majority of people I encounter equally want to grow, learn, and prosper while committed to being uniquely themselves within relationships. As we work together towards common objectives, we strengthen the bonds of key relationships, fostering a collaborative spirit and a shared sense of achievement when goals are met.

I understand that getting to this point where relationships thrive, values are shared, and objectives aligned is no small feat for any individual and even much less among so many in a corporation. I also understand that there are so many factors (equity, fair compensation, more) involved in creating successful relationships where each individual can exchange values without fear. But I also know that the effort is worth it, and the dividends will pay out over the life of a career (and even post-career) when given the proper attention and care.

Emphasizing communication.

I have felt the rewards of establishing open and honest communication and also the pain when I've neglected doing so. Communication is the cornerstone of my most successful relationships; hence, the ability to express thoughts and feelings openly, and actively listen to each other, is essential for understanding and connection.

I earned one of my master's degrees in business management from George Washington University (Washington, DC). There, I recall a class in project management that illustrated the importance of communication and how the lines of communication can quickly become unmanageable. Not to date myself too much, but this was well before the explosion of social media platforms; yet there were still many ways a message could get tangled. In the corporate setting, streamlining communication is a goal, but sometimes that's impossible when you have to work across many teams, stakeholders, clients, and even communication mediums. Whenever I initiate communications, I think of how quickly these multiple factors can cause just one message to get out of control, unmanageable, and thus ineffective. If in doubt, I think about whether my communication will allow colleagues to work through challenges and conflicts in a constructive manner.

Trust above all.

I once underestimated the effort necessary for building trust, which I now know takes time and consistency, honesty, reliability, and integrity. When I heard the word in my earlier years, I had a certain naivete about trust. Bright-eyed and highly optimistic about life, until that point success was pretty much a linear path with no twists and turns. Go to school, get great grades, work hard, and opportunities emerge. I figured that if I merely continue to pursue and focus on emerging technologies, STEM, education, and training, all would be well. And for a lot of my career, when I put in A+B, I received the expected C+D. Phenomenal.

However, I started to realize that, even as I was putting in more of the same ingredients, I was not getting the same outcomes. That's when I stumbled upon the hidden, unmeasurable mystique of trust. Trust shows up in situation upon situation, in life event upon life event. But trust varies in infinite ways. Someone might trust you to run a meeting with executives, but they may not trust you to prepare the income statement for the same executives. Someone might trust me to speak to potential investors but not to steer a meeting of engineers trying to improve product design.

Trust is also timebound. Experience can both buy you into a trusting relationship and at the same time expel you from one. It's now important to me that I recognize who to trust and at the same time who trusts me. Trust is essential for healthy and successful relationships, creating a sense of security and safety for me. In broader context, trust allows partners, friends (and even pets) to be vulnerable and open with each other.

Respect and empathy.
Respecting others' opinions, boundaries, and differences is vital for thriving relationships. Valuing and appreciating the unique qualities of others fosters a sense of mutual admiration and understanding. Learning and respecting boundaries and individuality contributes to a healthy and balanced relationship, creating an environment of acceptance and support. When I exude empathy, I can connect on a deeper level, engendering emotional intimacy and closeness within the relationship. My pursuit of empathy also helps me expand my ability to understand and acknowledge the feelings of others.

Dominique (Dom) Green helps position Fortune 100 companies for success as they adopt emerging technologies. He is the recipient of several academic scholarships and a *USA Today* Engineer of the Year Awardee. Dom has obtained undergraduate and graduate degrees in engineering, business management, and innovation technology. He is most passionate about helping at-risk communities.

Curating Your Brand: Community, Advocacy, Exposure

by Jasmine Grant-Hibler & Alex Smith

As a middle manager navigating the corporate landscape, Jasmine came to realize the paramount importance of curating her personal brand. In today's tech industry, where competition is fierce and opportunities abound, distinguishing oneself is not just beneficial but imperative for career advancement.

Alex has navigated this journey with a focus on community, advocacy, and exposure, recognizing the significance of these elements in shaping career. Alex's decision to attend Morehouse College was inspired by the legacy of great men like MLK, Spike Lee, and other notable civil rights leaders who walked its halls, instilling a deep sense of purpose and responsibility to carry their legacy forward.

In this essay, Jasmine delves into the significance of personal branding specifically tailored for African American senior leaders in technology. Understanding the value of one's brand extends beyond mere recognition; it shapes social networks, influences office politics, impacts calibration, and ultimately affects performance.

But first, Alex will address personal branding.

~~~

Alex: Early in my career at a major consulting firm, I had my sights set on a particular role, one I believed would allow me to make a significant impact in Diversity, Equity, and Inclusion (DE&I). However, that role wasn't viewed as highly within the company and was one of the first to be canceled during a downsizing. Initially, this was a setback, but I soon realized it may an opportunity in disguise.

Rather than being discouraged, I embraced this turn of events as a chance to redefine my path and make an even greater impact—a moment of clarity that shifted my perspective. I realized that my value to the company and the cause of DE&I wasn't limited to a
~~~

specific role or title. Instead, it was about how I could leverage my expertise and passion to effect change in any role I undertook.

As I transitioned into a new role, I found myself in a position of even greater influence because it allowed me the power to shape the composition of project teams, ensuring that they in fact were diverse and inclusive. The position enabled me to lead by example, demonstrating the importance of diversity and inclusion in every aspect of our work.

I not only managed project teams but also acted as a counselor, recruiter, and interviewer. These responsibilities were opportunities to make a positive impact, ensuring that our teams were not only diverse but also inclusive *and* equitable. By actively recruiting and hiring individuals from underrepresented backgrounds, I could help create a more inclusive workplace and drive better outcomes for our clients.

One of my key strategies was to actively seek out diverse perspectives and voices, making a conscious effort to include team members from a wide range of backgrounds and experiences, believing that diversity was essential to innovation and success. This approach enriched our team dynamics and led to more creative solutions and better client results.

I made it a priority to mentor and support junior team members, particularly those from underrepresented backgrounds, empowering the next generation of diverse leaders, who would create a more inclusive and equitable future workplace.

That job taught me valuable lessons about the importance of defining your brand and leveraging your expertise and passion to effect change. It showed me that true leadership is not about the title you hold or the role you play, but the impact you make and the values you embody.

Jasmine: To embark on this career journey effectively, it's essential to grasp the essence of personal branding, which is not about self-promotion but aligning one's values and skills with career objectives. I'll discuss how benchmarking one's brand against external groups can enhance exposure and foster community, advocacy, and exposure.

Defining your personal brand is a fundamental step in navigating effectively within the corporate landscape. By understanding what sets you apart and what drives you professionally, you can craft a brand identity that resonates authentically with others.

To begin, take stock of your values, passions, and strengths. Consider what motivates you and what you excel at professionally. This introspective process allows you to pinpoint the essence of your personal brand and articulate it with clarity.

Next, assess how your brand aligns with external groups and industry standards. Benchmarking your brand against these external metrics provides valuable insights into where you stand and opens avenues for exposure and growth. By engaging with relevant community organizations, whether through professional associations, online forums, or networking events, you showcase your brand and foster meaningful connections.

Moreover, leveraging advocacy and community engagement can amplify your brand's impact. By working alongside causes or initiatives that resonate with your values, you contribute to positive change while positioning yourself as a thought leader. Whether participating in diversity and inclusion initiatives or collaborating with industry organizations, advocating for meaningful causes can elevate your brand and expand your sphere of influence.

Defining your personal brand requires a deep understanding of yourself, your values, and your professional aspirations, aligning these elements and benchmarking against external standards, to enhance exposure, foster community, and ultimately propel your career forward with purpose and authenticity.

Leveraging community service via strategic volunteering. Choosing the right community organizations can be pivotal, amplifying your brand and building professional development. View strategic volunteering as another opportunity to grow and demonstrate your skills as a senior leader. Selecting which organizations reflect your brand, think about how your values line up with the mission and purpose of the organization. Are there opportunities for you to grow and develop new skills and demonstrate impact? Be clear in your own mind how these opportunities may bring recognition (with supporting metrics) to demonstrate achievement. Active involvement in non-profit leadership roles and board seats can also catalyze career growth opportunities.

Be sure your list will also let you amplify your passion. Are you a devoted advocate of supporting children for STEM education, single mothers looking for career opportunities, environmental causes, housing? The communities that bring together an intersection of your brand, passions, and opportunities for growth and development are the best investment of your personal time.

Once you have narrowed your list, focus on fit and commitment. Do they have leadership or board positions or only ad-hoc volunteering opportunities? Be realistic about how much time you can devote. Can you leverage your opportunity with the community organization to resolve development gaps or latent skill sets?

Alex: Here's an example of this. As a delivery manager in the technology space, I needed to expand my digital footprint. I engaged with a local nonprofit that worked with other nonprofits to build websites that included payment capabilities and common chat engines to provide a digital interface to their clientele. By signing up for an advisory role, I was able to learn, contribute, and eventually take credit for delivering digital capabilities. This was a skill set I needed to compliment my work resume and the volunteer opportunity \ was a great avenue to fill that skill gap. I was able to use that experience back in my performance cycle at work to demonstrate a new skillset and then capitalize on that experience.

Jasmine: As you embark on volunteer opportunities, make sure you translate the value of your experience. Put together a volunteer resume and ensure it quantifies your impact. For example: "I served as a senior tech advisor to the Boys and Girls of Atlanta and with an investment of $50K over six months we were able to build out a website with payment capabilities for donations and purchases as well as install a chat bot. The organization realized an increase of $500K in annual donations because of this investment in opening digital payment channels."

Lastly, determine the right channels to share that great work. It may be on your resume, in a biography, during performance review, or over social media (LinkedIn is a popular venue to share the intersection of professional and community achievements, where you can list involvements and post pictures of you in action with your community organization.) Take credit for what you put into the universe! Some useful websites to help you identify which community organization might be a good fit are Volunteermatch.org, Pointsoflight.org, Justserve.org, Unitedway.org, Boardsource.org, Idealist.org, and BoardnetUSA.org.

Advocacy in the Tech Industry.
As a middle or senior technology leader within the Fortune 500, advocacy is not just a responsibility but a powerful tool for driving positive change—especially engaging in diversity and inclusion initiatives. As a minority leader myself, I understand the importance of sharing our experiences and perspectives to foster a more inclusive environment. Affinity groups provide opportunities to connect with others who share similar backgrounds and challenges, collectively working towards improving diversity and inclusion within our organizations and beyond. Through these initiatives, we can help shape policies and practices.

Another key aspect of advocacy is collaborating with industry organizations. By identifying and partnering with organizations in our functional vertical, industry, or trade, we amplify our brand and contribute to larger industry-wide initiatives. These collaborations provide valuable networking opportunities, allowing us to leverage collective resources and expertise to address common challenges and drive innovation. Whether through sponsoring events, participating in working groups, or sharing best practices, we help shape the future of the tech industry while also enhancing our organization's reputation and impact.

Extracurricular exposure.
Extracurricular exposure can significantly enhance both personal brand recognition and industry influence. One effective strategy is to seek nominations in industry groups and awards. By positioning

ourselves for recognition in these forums, we accelerate our brand recognition and credibility within the industry. Awards and memberships in prestigious industry groups validate our expertise and can open doors to new opportunities for collaboration, mentorship, and professional growth.

Speaking engagements and panel discussions are another key avenue that allow us to demonstrate knowledge, executive presence, and what makes our brand unique. By sharing insights, experiences, and innovative ideas on relevant topics, we position ourselves as thought leaders and showcase our organization's capabilities and values. Presenting at industry conferences, participating in webinars, or speaking at university events raise our profile and contribute to the broader discourse.

Networking at conferences and events is key for staying engaged and strategically investing for career benefits and overall happiness. This involves being selective about which events to attend and actively nurturing relationships with peers, mentors, and potential collaborators. I prioritize attending conferences and events that are in line with my professional interests and goals, such as technology summits, leadership forums, and diversity and inclusion conferences. Engaging in meaningful conversations, exchanging ideas, and building genuine connections, I expand my network while gaining valuable insights, support, and opportunities for career advancement. Then, I make a concerted, pro-active, strategic effort to stay connected through follow-up emails, LinkedIn messages, and occasional coffee meetings. This benefits both my career trajectory and overall fulfillment.

Crafting a consistent narrative.

Crafting a consistent narrative is essential for building trust, credibility, and influence both within and outside the organization. One key aspect of this is matching personal values with professional endeavors, being purposeful about the corresponding level of commitment between our personal values and the organizations we represent, thus creating a win/win where contributions make a difference and are also personally fulfilling. Whether it's advocating for diversity and inclusion, championing sustainability initiatives, or promoting ethical business practices, integrating our values strengthens our authenticity and enhances our ability to drive positive change.

Another element of crafting a consistent narrative is showcasing expertise through our online presence. On platforms like LinkedIn, commonly used for professional networking, go beyond the basics, and showcase expertise and thought leadership—publishing articles, sharing insights, or participating in industry discussions. Our online presence should reflect our depth of knowledge, passion for innovation, and commitment to excellence. Even if certain platforms don't readily showcase expertise, we can curate and share content that highlights skills, achievements, and unique perspectives.

Managing our online reputation involves adhering to high-level best practices while leveraging digital channels for brand promotion and thought leadership. Some key dos are maintaining professionalism, engaging authentically, and regularly updating profiles with relevant information and accomplishments. On the other hand, don'ts include oversharing personal information, engaging in controversial or divisive topics, or neglecting to respond to inquiries or comments in a timely manner.

Case studies and success stories.

By examining success stories and case studies—those with good or poor results—we gain insights into the key strategies and pitfalls when shaping our own brand narratives. By studying success stories across various industries, backgrounds, and career trajectories, we can identify common elements, such as a clear definition of one's brand identity, encompassing values, strengths, and unique selling points.

Here are examples of technology leaders known for successfully curating personal brands.

- Ursula Burns, a trailblazing executive, served as CEO of Xerox from 2009 to 2016, the first Black woman to lead a Fortune 500 company. Known for strong leadership, strategic vision, and advocacy for diversity and inclusion in the tech industry. Burns effectively curated her personal brand by leveraging her experiences as a minority leader to champion initiatives aimed at fostering equality and representation within Xerox and beyond.

- Tristan Walker, founder and CEO of Walker & Company Brands, a consumer goods company focused on meeting the needs of people of color, is known for his entrepreneurial spirit, innovative products, and commitment to addressing diversity gaps in the consumer goods and technology industries. Walker has successfully curated his personal brand as a leader dedicated to empowering marginalized communities and driving meaningful change in the business world.
- Satya Nadella, CEO of Microsoft, is renowned for authentic leadership style and for aligning Microsoft's mission with his personal values of empathy and inclusivity. He has effectively communicated this vision through various channels, including public speeches, interviews, and his book *Hit Refresh: The Quest to Rediscover Microsoft's Soul and Imagine a Better Future for Everyone.*
- Sheryl Sandberg, COO of Facebook, is recognized for her strong personal brand centered around empowerment, leadership, and gender equality. Through her book *Lean In: Women, Work, and the Will to Lead*, and initiatives like LeanIn.Org, she has inspired countless individuals to pursue professional goals and advocate for diversity in the workplace.

Final words.

Jasmine: Curating one's personal brand is not a one-time endeavor but an ongoing process. By leveraging community, advocacy, and exposure, African American middle managers in the tech industry can carve out a distinct identity, propel career advancement, and foster a more inclusive workplace culture.

Alex: It took me a while to realize how I could embody the spirit of someone like, say, Spike Lee. I discovered that I didn't have to "brand myself" solely in DE&I space; instead, I could strive to be the best at my craft, add value, and drive business outcomes. I didn't have to be another MLK; I had to play my unique role, leveraging my expertise and passion to effect change in my own way. This realization transformed my approach to management development, guiding me towards a more holistic understanding of how I could make a meaningful impact in my career and beyond.

Jasmine Grant-Hibler, a seasoned professional with 25+ years of experience, spearheads the Treasury Solutions Technology division at Truist Financial Corporation. Her expertise spans various financial sectors, including treasury, broker dealer, private wealth, and mortgage. With an MBA from Goizueta Business School, she's a Women in Technology award recipient, passionate about global travel, motorsports, and community involvement. Formerly, she served on executive boards for the nonprofits The Atlanta Commission on Women and AIDAtlanta.

Alex Smith is an accomplished Strategy Senior Manager at Accenture, renowned for solving client challenges for Fortune 100 companies. With a portfolio spanning 50+ projects, he adds value by mapping out Next Gen solutions to drive digital transformations. Alex's professional passions include design thinking, storytelling, and shaping a culture of the future.

You are Your (Ever-Evolving) Brand

by Shanita Drake

Who are you? What are your values? Your unique characteristics? What sets you apart from others? What do others think of you—and how do you know? The answers to these questions are the beginning of discovering your brand.

I like to think of brand as how it's been described by many professional leaders: that it's what others say about who you are, when you are not present. Your brand serves as a representative; the culmination of characteristics that come to others' minds, the group of adjectives that become synonymous with the idea of working with you. Your brand is the essence of who you are and what you represent—the embodiment of what people think of when your name is spoken in a room your eyes have yet to see. If built successfully, your brand will be your best self-marketing tool, opening doors and bringing people and opportunities to you.

If I had known the value of building a brand as a young career professional, I would be light years ahead in enriching experiences and opportunities. As I entered corporate America, my idea of success was working hard and letting that work speak for itself. I was the first to raise my hand for a tough assignment, taking on extra work and working more hours to prove my dedication. It took years before I realized working hard is only a small part of the formula—and usually only rewards you with more work. As I've grown in my corporate career, I've learned that building the right brand is a quintessential part of one's career journey. It's not just about work ethic, but about how I teach others, how I handle adversity, how I solve problems. Am I likeable, approachable, and do others like working with me?

I can remember as an informal leader, receiving feedback that I was not inclusive. I had a team member who needed to know that I was interested in their personal life, so they'd feel a part of the team. Initially, I did not understand the need or behavior, however, after some coaching, I began to engage in a different way. As a result, we created a great rapport and that person's work performance changed for the better. Something so simple as asking "How was your weekend?" allowed me to build a better reputation as a leader and was more representative of the brand I desired.

Whether pursuing leadership or a new opportunity, your brand can either help open or quickly close a door. Oftentimes your brand will make an introduction before you even know people or opportunities exist. Rest assured, a brand is a component of your total package so you will surely want to dedicate time and attention to nurturing it with intentionality.

You have what it takes. It's not rocket science.

Growing up, I was not exposed to business culture or corporate speak. At home, there were no discussions of networking, brand building or awareness, no idea of what emotional intelligence was or having a personal board of directors. I was taught to work hard, be a good person and learn to adjust, and that my work ethic, skills, and ability would provide the exposure needed for career advancement.

Perhaps as a result, my career journey had no clear or linear path. I found myself moving from assignment to assignment or role to role based on interests or opportunities that presented themselves. I had just as many lateral moves as promotions and initially, my career was a series of passion moves over strategic moves. What fueled my excitement for forward progressions were childhood teachers who emphasized phrases like "Your attitude determines your altitude," and "Be the change you want to see," as well as "Treat others as you would want to be treated," and "You're young, gifted, and Black." These were the mantras I used to guide me through life, and I suppose you could say they formed the beginnings of what I would call my brand.

I have always been a student of people, discovering characteristics I wanted to emulate and those I did not. I'd spend time getting to know leaders I respected and admired, gravitating towards those who were people-oriented and giving, those who were successful and helpful and shared in the success with their teams. I asked questions about their career journey and how they got to where they were. I asked for their thoughts and opinions on current events and why they felt that way. I asked for advice from someone pursuing a similar path and I routinely followed up for additional insight or to offer my talent and time. It is in these moments that I shaped my visions of how I wanted to show up as a leader. It was in these very moments that I continued to build and modify a blueprint for my brand.

At this stage of my career, now that I have built a strong brand, I seek feedback to be sure it's currently what I want to project. According to feedback, I am known for my ability to execute and get things done. With each new assignment I take time to observe the environment as I create my plan of what I want to improve upon. I invest early in building relationships and asking questions. I build allies and ask for insight. The allows me to anticipate what's coming and plan for alternate solutions that helps make me a great solution provider. I understand that I must take some calculated risks, so I am comfortable operating in the gray. My prior call center experience has prepared me to take in information quickly and make decisions. As a technology leader in a hyper connected world, I am a collaborator and thought partner. I often meet with business partners from different segments of work just to learn and ask questions. This is how you build your network and expand your thinking. Sharing thoughts with others and being open to new ideas is a great way to gain exposure and reinforce the brand you desire.

Over time, time, through observation, consistency, openness, authenticity, and intentionality, I have built a brand that I am proud of. If you haven't done that yet, it is never too late to start. Consistently demonstrate who you are and what you value. See feedback as a gift. Have trusted advisors, mentors, and coaching—this allows me to keep a pulse on my brand and course correct when

I was not showing up the way I intended. Being a person of your word, adding value, and allowing the best parts of your character to shine will jumpstart your journey to managing your brand.

Your brand is your unwritten resume, and the truth is, you already have a brand whether you intentionally built one or not. The key is knowing what your brand is and managing it so that it exudes what you want. A well-managed brand can be the key to unlocking potential opportunities, career moves, and relationships, and grant you access to people that will help you navigate situations with grace and ease.

A strong brand builds confidence in others about your skills and ability, your values and character. When they know your brand, people feel that they know and trust *you*. People prefer to do business with those they like and trust, those with whom they share common values, people who are known for consistently operating in a certain manner.

Building a strong brand starts with meeting and exceeding expectations in your current role assignment and responsibilities, being dependable, reliable, and effective at producing results. Once you establish yourself as a solid performer, your brand will lead you to new experiences that should increase your exposure and visibility. An increase in exposure expands your network and provides even more opportunities. Leverage this new network. Make it a point to market what you do well and how that has helped advance the organization's mission and vision. As you continue to perform well and showcase accomplishments, you will begin to be recognized and acknowledged. Accolades and acknowledgements in turn boost your confidence and credibility. The more credible you become, the more expansive your sphere of influence.

Very early in my career as an analyst, I had responsibility for leading a team of individuals who were predominantly male and much older than myself, with more years of technical experience and no real incentive to listen to me. However, I was the one responsible for producing business outcomes. This is where my ability to bring multiple groups together, serve as a "translator," and produce business outcomes became known. I often referred to this strength as being bilingual—able to break complex technology issues into business terms so that both the business and technical teams could achieve success. The ability to flow from requirements to technical capabilities that solve customer issues is a vital skill. Understanding how to speak to the smallest components of a business application or process and help developers see a broader end-to-end picture of functionality for an end user was great for my career. Soon, I found myself with the ability to choose my next assignment or being placed on special assignments. All due to my brand.

Not only did I do my job well, but I also found a niche where very few were operating, and that in turn also became part of my brand. Via new opportunities, I increased my network, gained new skills, and met leaders who in the future would serve as mentors, advocates, and sponsors. You never know when someone is observing how you deal with difficult situations. Are you a collaborator? Are you driven? Are you clear, concise, and confident? At a critical moment, someone may speak on your behalf because they witnessed the way you get things done. With proper brand management, over time you can become a highly sought after individual.

Can I change my brand? Absolutely! As you grow and evolve, so should your brand. Just like the pristine engine of a high-powered sports car, your brand requires maintenance. As you progress through your career you will learn new things. Through life experience, you may even change your viewpoint or position on topics that you once held a firm position on. Then, you incorporate those changes into your evolving brand.

Remember my story about the team member who did not feel included? Imagine if I were only given that one chance to show that I was inclusive. That would have been career limiting. Your brand is not just one good or bad experience, but a series of consistent interactions. If your brand is not representing what you desire, you can always work to change the narrative. Even if your brand is great but you want a different brand, you can start revising. It is all a part of the journey, ad the beauty is, we are all allowed to change. In fact,

I encourage you to change. As career professionals we are expected to learn and become better. The early career version of you may be vastly different than the middle career or later career version of yourself. Your ideals, values, goals, communication style, and preferences will all evolve. You will realize things about your character that you no longer desire or do not serve you well.

It could take years to build a brand you a proud of and only a few unfortunate encounters to diminish your hard work. Self-awareness will help you maintain the brand you desire. Understanding how you show up both in the physical world and virtually is important. Understanding how others experience you is key to a realistic sense of what your brand is versus your perceived (or hopeful) view.

Tips for fine-tuning your brand.

- Feedback is your friend. Seek feedback annually and 90 days after taking any new role or assignment. Ask open-ended questions that allow others to describe their observations.
- Evaluate your evolving value system. What do you have a new perspective on or what are newly established values?
- Review your social media (posts and tags). Ensure they represent what you want them to, and if not, delete as best you can.
- Periodically refresh your network. Review personal contacts, reach out, connect, reconnect, or disconnect.
- Be authentic and consistent.
- Remember YOU are your brand!

Shanita Drake is a dynamic Technology Director with an impressive 25-years of experience in the industry. Her ability to align technology strategies with business objectives has consistently resulted in streamlined operations, enhanced productivity, and increased profitability. With an unwavering dedication to excellence, she fosters a ollaborative work environment and drives high-performance results. Shanita's passion for family and ommunity is evident in her coommitment to serving others. She is guided by the principles of setting goals, overcoming obstacles, and spiring others see the potential in every situation.

Part IV

Crashing the C-Suite: Setting Your Place at the Table

"A cat that dreams of becoming a lion must begin with losing an appetite for rats."

- African proverb

Constructing a Path to C-Suite

by Troy Wise

I never had a plan to achieve a certain level of leadership in any organization. My original career goal was to be an autonomous business owner. You see, my early role models were my grandfather and uncle who each had their own businesses—auto repair and irrigation and lawn care. I watched them work hard to grow their businesses and provide jobs to folks in the community. However, they worked long hours, and the physical strain over time took a toll on their bodies. I wanted to find a way to work smarter by maximizing my earning potential while having a great quality of life.

But their influence was strong, so much so, I started working at 14 years-old, having several jobs and side hustles throughout high school. I was so committed to working that I didn't think college was for me. Then the real world hit. I wanted things and the rate I was making money, whether working for others or running my own cleaning business, wasn't keeping pace with my desires. Fast forward to graduating with my MBA, faced with the choice of continuing business ownership, or entering a corporate career. I decided to pursue both; thinking at some point I would make enough connections to eventually focus solely on growing my business.

Call it lack of planning or lack of depth in planning. My corporate career became my primary focus, but I still struggled with imposter syndrome. Although highly educated, I never felt completely comfortable with senior leaders of the organizations where I worked. Then, one fateful day I got the boost I needed from a very unlikely source.

I was at a conference hosted by a national vendor of large-scale document processing machines. I had developed a plan to automate all customer letter generation and we needed equipment to manage the additional volume. At the conference, I met and observed many senior and highly credentialed IT leaders. Still relatively early in my career, I felt less than qualified to make such an important decision for the firm. I attended the conference with the head of purchasing, and in a moment of insecurity, I leaned over to him and said, "I don't belong here."

He looked at me with an intense stare and said something that struck me as some of the most significant words I had ever heard: "No-one in this building is more important than you. No-one!"

The statement moved me like I had just won an Oscar, Grammy, or the Super Bowl. He spoke confidence into me that I had never felt before. I wondered how this older Jewish man can have more faith in me then I have in myself. In that moment, I swore that I would never doubt myself ever again. Why would I? I knew my work and study ethics were top notch. It was instilled in me at a young age to never let anyone outwork me, in a job or sports. I had also learned through my undergrad and graduate school experience that your background doesn't define your success, but your ability to find your style of learning and persistence will move you to the top of the class.

With this newfound confidence, I completed the assessment of the top products and submitted my recommendations to the CEO and CIO. They proceeded, without question, to purchase the equipment I recommended. My team implemented the new software solution, and we exceeded the letter generation volume forecast with a much higher rate quality than expected. The success of that effort, and my confidence boost, resulted in my being assigned more complex business problems, a promotion, and a larger team.

As I progressed in my career and moved from role to role and from company to company, I realized I needed to not only focus on building new connections but also establishing relationships that stood over time and as I changed companies. I had to embrace a group of sponsors and mentors. I wasn't intentional about any goals of the relationships; I just wanted a group of people I trusted and respected to bounce ideas off. Most were former bosses who I would connect with from time to time over coffee or lunch. We would talk about sports, family, careers, and company dynamics. One day I was listening to a senior executive discuss their career journey and she mentioned the importance of what she labelled as a personal board of directors; that just like any successful business, I should have an established set of advisors to provide counsel as needed to ensure success. Many books and training on career coaching call the members of that group sponsors, mentors, and allies. Although I never considered my relationships as my own board of directors, it made total sense to me.

Thereafter, I leveraged my personal board of directors whenever I had a challenge to solve and needed guidance.

A few years ago, I was working for a very difficult CTO, causing me to doubt my capabilities. It was a very tough time for me. While trying to figure out a working relationship with this new manager, I had to process why my career might be stalling. I felt (and was told by many colleagues and leaders in the company), that I should have his job. Perhaps I had been the top internal candidate, but the firm wanted someone with outside experience in the domain. With that decision, it seemed I hit a wall in my career. I had been in my role for four years, when I was used to moving out of roles every two to three years on average.

I had to take a moment for self-reflection: Am I in the right organization? The right company? Had I reached the ceiling based on my level of competence? Have I done enough networking? I was well known as a high performer, but it seemed I was not well positioned for the next level.

I decided to check in with my board of directors and met with a few; one of the most revealing, epiphany-like conversations occurred with the head of emerging segments. She had me walk through my elevator pitch, the stories I told in interviews, and offered a bit of advice that I now pass on to others. She advised me to embrace my accomplishments and share them, but to relate them to business outcomes. Go beyond just completing the software delivery. What is the value you provided to the company or to our customers? Did you reduce risk or expenses? Did your work increase revenue? Did you improve customer or team-member confidence in the brand?

When I surveyed other members of my board of directors, I realized they all were very successful at telling a story that connected to business value, goals, and customers. I sought new material to read and new trainings. As a result, I realized that I had a talent and skill to solve complex business problems that others were afraid to touch. I could frame business needs to drive comprehensive solutions, and then position those solutions to drive the expected outcomes and simplify technical complexity in translating roadmaps and execution plans to business leaders.

First, to offset any challenges I was facing with my current manager, I had to lean into knowing I was a beast at what I do. My board members suggested that I always have a strong relationship with my business partners and my second level manager, my boss's boss. My connections with business partners were already strong but I had to take action to form a relationship with my next level manager, who was new to the organization. Over about six months, I built a decent relationship with my boss's boss, and we discussed mobility, business pain points, and direction of the organization. I was then pulled from under my manager into a special assignment driving transformation for all of consumer technology and partnering with two others to drive transformation for the entire technology organization. Although the work was great, I had to step out on faith. With this move, I had to give up the team I built that had grown from barely delivering and struggling, to building trust with business partners, and eventually into a high performing software delivery team that had a seat at the business table with great influence over product direction.

With stronger confidence in my abilities to bring people together and to solve large and complex problems, I crushed the assignment and gained a new sponsor in my boss's boss, who then created a role for me and became my direct manager. That elevated me to peer level with the manager I had struggled under. Our relationship shifted. That difficult manager now recognized some areas of opportunities, while acknowledging my ability to deliver in a way that people loved to follow. He stated that once I become more courageous with making hard decisions quickly, I would supersede him and excel to levels in the organization that he would not reach—a satisfying compliment.

By leveraging the knowledge of my board of directors and the deep understanding they had of me, I was able to position myself to overcome a difficult manager-employee relationship, while breaking through the stagnation in my career. They continue to help me build confidence to seek positions I never imagined having in the corporate world.

Troy Wise is a Technology Executive at Wells Fargo and serves as Head of Shared Services in Consumer Technology. He is responsible for delivering services across several technology areas, consumer intelligence automation, credit decision-making, and data. Troy has over 25 years of experience leading technology organizations in financial services and healthcare. He has been with Wells Fargo for 18 years serving technology roles across commercial and consumer lending. Troy lives in Maryland and enjoys cis, and spending time with family.

Understanding the C-Suite Ecosystem and Terrain

by Messay Abebe

Navigating the culture.

I have always assumed that the makeup of the final step in the corporate ladder, the C-Suite, is the Chief Executive Officer (CEO) surrounded by other executives and senior managing directors. But the reality is that the C-Suite is much larger, with dynamic leadership roles structured on the intrinsic value proposition of the corporation. Chief Executive Officer (CEO), Chief Information Officer (CIO), Chief Operating Officer (COO), Chief Technology Officer (CTO), Chief Marketing Officer (CMO), and Chief Human Resource Officer (CHRO) are all C-Suite roles—and their organizational culture can create a challenge if you don't understand how to navigate their interdependencies.

In my career, I had a couple of opportunities to observe strategic forums led by two separate Fortune 500 C-Suite executives. In those presentations, although the core tenants of the organization's value proposition did not change, to see the dynamic cultural differences in decisions making was glaring. I was enamored by the intensity, its fact-based dialogue and debate, forcing discourse on what I sensed were even opposite ideas to verify the merit of final decisions. There was no doubt of the clear power structure in the room. I also noticed that despite differences in these two separate C-Suite executive strategic sessions, each exhibited the corporate values of inclusion and the idea and trust in teamwork.

Collaboration is essential.

Reaching the upper echelon of corporate leadership requires years of execution, effective leadership, and savvy navigation/networking of corporate politics. For African American leaders, the road can be steep; you need to recognize the power of collaboration early on and commit to engaging in the fierce political/networking required to succeed. I have observed many operational all-hands forums where C-Suite members shared visions and leveraged collaboration to build on their peers' strategies. While there was a level of inherent competition (each leader focused on highlighting their organization's success), it was clear that shaping the company's direction and culture required close alignment and coordination. Moments like these taught me that the game does not end when you reach the C-Suite. Being a part of that exclusive club requires constant networking, collaboration, and political maneuvering. Getting a seat at the C-Suite table is a process of fierce competition, but being successful in that seat requires understanding the overall culture and collaborating to drive strategic direction.

Interconnectivity and intersectionality of success.

As I gained more knowledge of the horizontal C-Suite executives' operational mandates, the overlapping accountabilities each leader held within their divisions became obvious. I recall a training I undertook as a technology executive where our CIO provided the vision of our financial institution's technology strategy to transform digital banking. Soon after he concluded, his peer (the CTO) shared the firm's multi-year transformational vision to pivot from traditional banking operations. Each presentation was confident and well crafted, but I was stuck by their interdependencies. I knew then that the success and competitiveness of our financial institution would be dependent solely on how these two dynamic leaders understood their common goals, and the interconnectivity of their mission and metrics. For one to succeed, they both had to succeed.

Mastering political dynamics and unraveling corporate politics.

Every business, small or large, has an operational culture that aids its success—and sometimes can factor into its demise. From the inception of the business, the culture comes alive, amplified by its leaders, and driven to team members and operations. In my 25+ years navigating industries ranging from government to consulting to major financial institutions, I was committed to understanding each organization's structure, and the operational culture. I saw it as a game, a game one must learn and navigate, gain advocates, and position oneself on the best path to success.

Yes, I am oversimplifying. Yes, corporations have clearly defined operational structures each with a power hierarchy and decision accountability. But there are also informal political structures where leaders have built strategic alliances through personal relationships or mutual operational advantages. The difficulty of navigating this complexity is exaggerated for minority leaders who may not have the ease to build personal relationships with Caucasian counterparts. I have always believed that recognizing this fact while also seeing it through a lens of "game" will assist you in quieting the whispers in your head that could throw you off from your potential and could effectively influence the favorable decisions you seek.

Cultivating political savvy.

Experience results from wins and losses in the corporate arena, from networks and alliances, and from your grit to excel. Throughout, you build and strengthen your acumen and political street smarts, learning to navigate the complex corporate hierarchy, thereby elevating your voice with merit and authority. Make no mistake, authority can be given but experience will determine the level of expertise and acumen you bring, which is closely collaborated with your ability to successfully influence and lead.

The power frameworks: Mapping the power structure for C-Suite Black executives.

Earlier in my career, the chief human resources officer at a Fortune 500 company invited me to join a C-Suite forum; she shared that, "knowledge of these formal and informal power networks is crucial" and invited me to observe the power dynamics. It's been 15 years since my first exposure to the C-Suite, and since then, I have enjoyed roles of growing responsibility across consumer products, telecommunications, industrial coatings, and financial services. I have since been afforded additional exposure to the C-Suite and in every role, I take, I make it a point to understand the composition of the C-Suite and the spheres of influence in the company, to always understand the flow of power. When I had my first opportunity to be a silent observer in a board meeting led by the CEO, I saw that while the board had key oversight function, it was still top management—the Chiefs reporting to the CEO were responsible for the company's performance and key decision making. I also noticed that the chain of command was known to everyone. Black board members and Black chiefs that I spoke with had a strong understanding of the power structure in the room, and they had learned how to use that power to their advantage.

Navigating with integrity.

Navigating the corporate ladder grows ever more difficult because performance at the C-Suite level is no longer about executing well, especially on the upper rungs. Excellence in execution is foundational, but it will not be the sole driver of upward movement. You must learn your environment, your alliances, your relevant spheres of influence, and at times you will be asked to make decisions that could question or undermine your values. As an underrepresented minority, you may feel the extensive pressure of an outsider and it may be tempting to compromise to continue climbing. Don't do it. Maintain your values and integrity in each decision. That quality becomes part of your character as a leader and will define you.

Messay D. Abebe is a Technology Executive at Bank of America leading critical integration and data service platforms. He brings 20+ years of software and platform resiliency solutions engineered to improve the financial needs of people.

Share Your Superpowers! (Yes, You've Got Some.)

by Joy Middleton-Saulny

Discovering your superpower.
I recently spoke at the Grace Hopper Celebration of Women in Computing, on the importance of institutions viewing privacy as a product in their organization. First, I read the audience and realized that I needed to alter my messaging to appeal to the varied levels of education and experience in the room; this seemed to strengthen engagement. After the panel discussion, there were many questions from the large, diverse audience and a true sense of engagement on a topic that, let's face it, is critical but not the most riveting.

My ability to consider my audience, adjust my approach, and deliver my message, while connecting authentically, all in the moment, is something I attribute to a very particular ability.

I am an engineer by training, with skillsets that come naturally to engineers such as systematic problem-solving, design, innovation, and more. I am a Black engineer, and some interpersonal skills developed out of cultural and survival needs come into play as well—collaboration, questioning for discovery, driving from ideation to the desired end results. As a Black corporate executive, I bring a unique perspective to the table, and because of my upbringing I understand and appreciate different cultures and backgrounds. This cultural intelligence helps me navigate diverse work environments and build strong relationships with colleagues.

These skills were well honed over decades. I recall Sabina Ewing, CIO of Abbott Labs, speaking of being a "corporate athlete" and that Black corporate athletes play to win versus playing not to lose. I believe that most Black corporate athletes concatenate all the skills in their tool kit and apply them as needed in every game; playing to win instead of playing it safe and all while knowing that failure or loss is not an option.

I discovered that my superpower as a corporate athlete was not only challenging myself to be the best player that I could be, but also serving as team captain. I knew how to leverage the strengths of everyone on the team, to design a new tool, develop a new and innovative patent idea, or transform a business. Like many Black corporate executives and C-Suite leaders, I possess strong leadership skills, enabling me to inspire and guide my organization toward achieving common goals. I do this by communicating my vision for the future, making difficult but rational decisions, and leading by example.

Networking.
How did you build a network of people who assisted in developing your power? As a Black corporate executive, I understood the importance of building a strong network of relationships that I regularly fuel. I work to connect with others, build trust, and create opportunities and drive business growth. My network helps me discern how to bring clear and tangible value to the organization. Black corporate executives face very specific challenges and obstacles; resilience allows us to overcome adversity and push forward in the face of setbacks. We bounce back from failure, and I have had my share, and learn, by having key, trusted advisors to help us think through and internalize the learning opportunities. When we can, we should also serve in that role for others. It's important to select advisors who share my values so that the lessons learned are in concert with deeply held beliefs.

Both advisors and mentors play a crucial role in career development and the success of Black executives. Here are some reasons why they are valuable.

Guidance, support, and honest feedback helps us navigate the challenges and complexities of our careers. Ten years into my career I self-selected out of an executive role to support my only child during a difficult divorce. I don't regret any moment of my decision; my daughter is now thriving in law school. The path back to an executive position after several years was difficult and I could not have navigated without trusted advisors and mentors.

Networking opportunities are frequently provided by advisors and mentors, who often have an extensive network and can introduce Black executives to influential individuals in their industry.

Skill development is ongoing. Advisors can discuss the key improvement opportunities with us and offer constructive feedback focused on growth and development, helping us navigate blind spots. Confidence building happens organically when surrounded by key resources and leaders who believe in you, boosting your self-confidence and helping fuel continued success.

The zones.
How do you move from the "zone of excellence"—the things that you have demonstrated you can achieve repeatedly over many years—to the "zone of genius," achieving that new level of success that you were not aware you have in you?

There is always a focus on our known superpowers. Doing the things that we do extremely well repeatedly and often. Our constituents constantly pull us into that zone, asking us to repeatedly deliver on known abilities. They need us in the zone. The problem with staying in that zone is that it does not allow growth and development. I recall working at Bank of America as a process engineering executive leading large scale process design and process improvement projects and teams. All my projects were successful, delivering improvements in efficiency and/or significantly reducing costs. In each case, I was using or polishing skills that were already in my tool kit. I needed to move to a program or an innovative project that was a stretch.

I spoke with my sponsor, and he helped me to take on a $40M project aimed at liability risk management for the bank—a project that let me learn a new area of the bank and deliver value to the bottom line. After that, I was committed to only taking roles that included a stretch.

The zone of excellence is comfortable, but the zone of genius unearths your true possibilities.

Each one, teach one. Or many!

Black executives need to unapologetically provide education, exposure, and/or promotions to Black up-and-coming professionals. Be cognizant of helping to pull others up the ladder. Remember the people who helped you. Understand your obligations to reach down, lend a hand to pull someone forward. My Ghanaians friends refer to this as Sankofa, meaning "looking forward as well as reaching back."

FAILING: Finding an Important Lesson, Inviting Needed Growth

Yes, let's learn from the past but not dwell on it; that energy could be used for better purposes.

We all have failures in our careers, and it's critical to analyze them. It's difficult to pick one failure that I learned the most from. When I was in banking operations serving as the southeast region executive, a new global leader of transaction services reporting into the technology and operations C-Suite executive came to the bank. She had a reputation as a tough and demanding leader who liked to surround herself with only men on her leadership team. She quickly reorganized and, despite my track record of delivery, moved me into a supporting role, and as my replacement, selected a male who I judged was doomed to fail. I was furious but tried to take one for the team. I walked into the office of the T&O C-Suite executive, a Black man, seeking guidance and instead I got moved to yet another segment of the organization and was blackballed. (Many males selected for that executive's leadership team knew that my replacement was a bad choice; they were just afraid to confront her with the truth.) Less than six months later the man who replaced me in the role was terminated for making racially and sexually offensive comments during his all-hands meeting addressing hundreds of employees.

Key lessons learned: (1) don't assume that a Black C-Suite executive is going to support you; make sure that you have an ally; and (2) sometimes it's important to be still and let things play out, no matter how difficult. There is always opportunity in failing. Find and internalize the lessons in needed growth.

C-Suite characteristics.

High-Level Decision Making. Responsible for critical decisions that shape an organization's strategic direction and future, C-Suite executives analyze complex information, assess risks, and determine best courses of action to drive company success.

Long Hours and Intense Workload. C-Suite executives typically work far more than 9-to-5. Demanding workloads require juggling multiple responsibilities, meetings, presentations, and high-level discussions.

Travel and Global Engagement. C-Suite executives typically must visit different locations to meet with clients, attend conferences, and establish partnerships to expand the organization's reach and make important connections.

Networking and Relationship Building. Relationships are a crucial aspect of the C-Suite lifestyle. Executives engage with other influential leaders, stakeholders, and potential partners to build and nourish those connections.

Constant Learning and Skill Development. C-Suite executives are continuously growing, developing skills, staying informed about industry trends, technological advancements, and market dynamics to make informed decisions and adapt to changing business landscapes.

High-Level Stress and Pressure. A highly stressful and demanding lifestyle is common for C-Suite occupants. Executives face immense pressure and must effectively manage stress and find ways to maintain their well-being.

Public Image and Reputation. Often the public face of the organization, C-Suite residents must represent the company's values, mission, and vision, and are responsible for maintaining a positive public image. This may include media interviews, speaking engagements, and public events.

Financial Rewards. C-Suite executives are typically well-compensated, receiving competitive salaries, bonuses, and other financial incentives that reflect their level of responsibility and the value they bring to the organization.

Work-Life Integration. Balance can be challenging for C-Suite executives who juggle professional responsibilities with personal commitments and self-care.

Impact and Legacy. Executives in the C-Suite can make a significant impact on an organization, its employees, and stakeholders. They have the power to shape the company's culture, drive innovation, and leave a lasting legacy.

Next, the boardroom.
Moving from a C-Suite executive role to the boardroom can be an important career move for Black executives. Here are some steps and considerations to help Black C-Suite executives navigate this transition.

Build a Strong Network. Networking is crucial for accessing board opportunities. Connect with board members, industry leaders, and influencers who can provide guidance and recommend you for board positions. Join professional organizations, attend industry events, and actively expand your network.

Develop Board-Ready Skills. Identify areas where you can enhance your skill set to align with boardroom requirements. This may include financial literacy, strategic thinking, risk management, governance, and industry knowledge. Seek opportunities to gain more experience in these areas through training programs, executive education, or advisory roles.

Seek Board Mentors and Advisors. Connect with experienced board members who can provide guidance and mentorship, offer insights on navigating the boardroom landscape, share experiences, and provide valuable advice on positioning yourself for board opportunities.

Pursue Nonprofit and Advisory Board Roles. Nonprofit organizations and advisory boards provide valuable board experience and help establish a track record of board service. Seek board opportunities or offer your expertise as an advisor to organizations in your industry. This can demonstrate your commitment to board service and enhance your credibility.

Leverage Executive Search Firms and Board Networks. Executive search firms and board networks can be valuable resources. Register with reputable board databases and engage with search firms specializing in board placement; they can help match your skills and experience with organizations seeking diverse board members.

Engage in Thought Leadership. Increase your visibility and credibility by sharing your expertise and insights by writing articles, speaking at conferences, or participating in panel discussions to showcase your knowledge and establish yourself as a thought leader. This enhances your reputation and makes you more attractive to board opportunities.

Seek Board Diversity Initiatives. Many organizations are actively seeking to diversify their boards. Research companies and initiatives that prioritize board diversity and are actively seeking out qualified Black candidates. Engaging with these organizations can increase your chances of being considered.

Network with Current Board Members. Actively network with current board members, including those from diverse backgrounds. Attend board events and industry conferences where you can connect with board members and gain insights into the boardroom dynamics. Building these relationships increases your visibility and opens doors to future board opportunities.

Develop a Board Resume and Online Presence. A board-focused resume that highlights relevant skills, experience, and accomplishments is key, as is establishing a professional online presence, including a LinkedIn profile that showcases your board aspirations, accomplishments, and thought leadership.

Engage with Board Readiness Programs. Board readiness programs provide valuable insights into boardroom dynamics, governance, and responsibilities of board members. They also offer networking opportunities with experienced board members and industry leaders.

Finally…
As I continue to hone my superpowers with the help of family, friends, trusted advisors, and mentors, my journey unfolds and perhaps will lead to a rewarding C-Suite career followed by a seat at the table in the boardroom! The most important role that I would have in both positions would be to create a legacy that makes it clear that I, and many like me, bring unique skills to the table which translates into value for the broader organization.

Joy Middleton-Saulny is a technology executive with nearly 25 years of experience in industries including manufacturing, telecommunications, industrial coatings, and financial services. Currently at Truist Financial Corporation, Ms. Middleton-Saulny is deploying digital tools to enable consumer trust and privacy. As an executive on the Bank Policy Institute and the Future of Privacy Forum, she works to transform data privacy practices. A 2022 Herndon/NASDAQ Director's Institute Fellow, Joy holds an MBA in Operations Management and a BS in Manufacturing Engineering from Boston University.

Influence and Allyship: Speak Up, Reach Out

by Sophia Ferguson

Alliance building inside and outside the C-Suite.

In my interactions over the last decade with minority C-Suite leaders, all indicated that the path is marked with unique challenges and experiences, made less difficult and more rewarding with solid alliances and allyship. The act of cultivating meaningful, heartfelt alliances goes beyond merely gaining entry into the C-Suite; it is essential for thriving and surviving within the executive ranks. Alliances, coupled with active allyship, unlock doors that may initially appear dauntingly closed. At the heart of every organization are complex human beings, necessitating alliances and allyship transcends backgrounds, race, ethnicity, and gender—so not just a necessity for minorities but for anyone looking for success in this arena.

Let's talk about alliances first. Just as strategic alliances are crucial for a nation, forming strategic alliances is a fundamental building block for navigating the complexities of the C-Suite and senior leadership in any organization. They provide insights into unspoken rules and power dynamics.

At their core, alliances are founded on strong, healthy relationships which demand a significant commitment of time and emotional investment to foster and develop. They need to extend beyond the confines of the C-Suite to yield meaningful results and impacts. The relationships you developed throughout your career are the foundation for forming these alliances.

In my leadership roles at Deloitte, I have realized that relationships between CEO, CFO, COO, CIO, and other C-Suite executives drive the cohesive strategy and operations of the business to make Deloitte the number one professional services firm. The effectiveness of the C-Suite is not merely a sum of the individual contributions of folks in the ranks, but a tapestry of interwoven relationships and alliances. As Winston Churchill said, "There is only thing worse than fighting with allies, and that is fighting without them."

Now, about allyship: The word resonated across the world in 2021, highlighting the need for change following several social injustices like the murder of George Floyd. It signifies a commitment as profound as the ocean—using one's privilege to uplift, to champion the cause of those who face systemic barriers. In the corporate sphere, this translates to a crusade against inequality, a fight to open doors, to build bridges where walls once stood.

The synergy of mentorship, sponsorship, and a personal board.

In my professional journey, I've learned the role of mentors and sponsors cannot be overstated. In addition, I now have a personal board of directors, a group of trusted advisors intimately familiar with my strengths, weaknesses, and aspirations. This is the essence of mentorship, akin to a friendship that evolves over time, offering a two-way street where learning and teaching are reciprocal.

Ursula Burns (Former CEO of Xerox), the first Black woman to lead a Fortune 500 company, often spoke about mentorship and sponsorship in her career, highlighting how those relationships were crucial for her growth.

The call for innovative leadership demands we become catalysts for change.

Steve Jobs once said, "Innovation distinguishes between a leader and a follower." This quote has profoundly influenced my career, highlighting the necessity for leadership that pushes past boundaries and embraces unconventional ideas.

Interacting with C-Suite leaders, I've recognized that innovation isn't solely about individual creation, but cultivating a space where creativity can thrive, allowing talented individuals to excel, resulting in the creation of exceptional products or services. The essence of C-Suite leadership lies not in single-handedly innovating but in building and empowering teams equipped for groundbreaking innovation.

Early in my career, as a solutions manager for finance and HR Systems, I came to this conclusion: Innovation begins with leadership that dares to be bold. I stepped into my role and very quickly identified a significant inefficiency—the IT team's week-long process to support the finance department's month-end tasks delayed crucial financial operations. Despite the team's belief that ERP software limitations, as corroborated by vendors and consultants, prevented automation, I was not convinced.

My curiosity and background in programming propelled me to delve deeper. Contrary to expectations, I developed an automated solution that reduced task completion time from one week to a mere 30 minutes. This achievement transcended technical skill; it was a testament to my commitment to challenging the norm. As a newcomer, I did not hesitate to question long-standing processes, driven by a quest for efficiency and a personal ethos prioritizing work-life balance. I was not about to take on tasks that would require me to work late evenings and weekends to meet deadlines! I wanted the time to get through my responsibilities so that I could build and pour into my relationships with people.

My resolve to challenge the conventional approach was initially met with skepticism, especially from my boss. I was cautioned to slow down and advised that this was a bold step towards transformative change; despite skepticism I pushed forward, choosing leadership and innovation over compliance. This decision not only propelled operational efficiency, but also established a new standard for problem-solving and creativity within the team. Without even knowing it, I opened the door to other innovative thinkers. It was a defining moment in my career which underscored the power of innovative leadership in transforming standard practices and fostering a culture of continual improvement.

When I reflect on that experience, three themes come to mind:

Your people are your inventors. The realization that the most critical assets for innovation are the people within the organization is paramount. Their varied perspectives, skills, and experiences form the foundation for innovative breakthroughs. Innovation starts with people who are encouraged to think differently. Embrace diversity and foster a culture of curiosity, where questioning and challenging the norm are encouraged.

Customer centricity is the engine of innovation. Gaining a deep understanding and foresight of customer needs is essential. Immersing oneself in the culture and context in which you operate fosters a more intuitive approach to customer-centric innovation. Devise strategies that keep you closely connected to customer preferences and market trends. This ensures that your innovative efforts are not only relevant but will make a significant impact in the market.

Strategic emphasis on technology. It's crucial to adopt a selective approach to innovation, rather than attempting to spearhead every emerging technological trend. It's more effective to concentrate on areas that align with your strategic goals and expertise. This focused approach ensures that your efforts in technology innovation are both efficient and impactful, driving forward areas that matter most to your organization. Remember, the role of the C-Suite is not only to envision the future but to create it, harnessing the collective power of your people, understanding customers, and intelligently navigating the ever-changing technological landscape to know where to focus.

Embracing the role of change agent.

In our fast-paced world of technology, embodying the role of a change agent goes beyond adapting to being at the forefront of innovation. My experience in the biopharmaceutical industry serves as a prime example of such forward-thinking leadership.

Upon entry into biopharmaceutical technology, I identified an essential opportunity: the integration of Artificial Intelligence (AI) and Machine Learning (ML) in healthcare. Noticing the industry's sluggish digital transformation, I envisioned a radical change in patient care through technology, informed by my in-depth research on the needs of healthcare providers and on patients' experiences. Capitalizing on that insight, I proposed a project to integrate AI and ML into patient care, to improve the speed to gaining prior authorizations so patients could get specialized drugs faster. I created my proposal (backed with compelling research and case studies) and aligned it with the organization’s mission to lead in healthcare innovation. I knew that I would receive pushback, as the culture had not embraced AI because of concerns around violating HIPAA regulations (a concern not grounded in facts), but I also knew it was the right thing to push for this change and innovation.

While I faced skepticism, I maintained my stance, and through workshops and expert-led seminars, shifted the narrative, highlighting AI's benefits and ethical applications. My resilience gradually turned skepticism into support. My leadership and tenacity in pushing for change resulted in successfully embedding AI within our healthcare systems, setting new standards for technological innovation in the industry. This breakthrough not only enhanced patient care but also marked our company as a pioneer in healthcare technology.

Clear vision and assertive leadership can drive significant advancements, setting new industry benchmarks. I encourage you all to embrace the power of your distinct perspectives. They are not just unique facets of your identity but catalysts for progress and innovation.

If I can see you, I can be you: emulating the experience.

There's something uniquely powerful about seeing a leader who mirrors your own background and dreams. It's a visual affirmation that says, "If I can see you, I can be you." For minority leaders like us, these words are a lifeline. Larry Quinlan's ascent within Deloitte wasn't just a professional milestone; it was a beacon of hope that shone directly into the heart of my aspirations. He and I share the same alma mater, the University of the West Indies. Deloitte was a top organization for me and as a matter of fact, I made my decision to work for Deloitte because Larry was the US and Global CIO.

I remember vividly the first time I heard Larry speak. His voice had a calm authority that filled the room, a presence so commanding yet inclusive it felt as though he was speaking directly to me. He shared stories of his journey from a humble childhood as the "skinny boy from St. Kitts" rising to become CIO of the world's largest professional services organization. Each word Larry spoke was threaded with the warmth of our shared Caribbean heritage, resonating in a place deep within me, stirring a mix of profound joy and tears. I was so moved that I couldn't keep this inspiration to myself; I shared his speeches with everyone who mattered in my life—from my university professors to the young minds I mentor—hoping his words would ignite in them the same boundless hope they sparked in me.

Larry's tenure as Global CIO was marked by groundbreaking initiatives and transformative strategies. He wasn't just managing systems; he was leading a global revolution in technology. His achievements at Deloitte were not just corporate victories. They were landmarks that showed what a kid from the Caribbean could aspire to and achieve.

In one of our encounters, his advice to me was simple yet profound: "Carry our legacy forward. Make us proud." More than mentorship, this was a call to action, an invitation to join him in breaking barriers and reshaping narratives. Every speech Larry gave, every panel he sat on, each decision he made at the helm, resonated with me. His journey was a live demonstration that passion, resilience, and authenticity could indeed carve a path to global leadership.

Larry's journey in the tech world at Deloitte wasn't just a career trajectory to admire; it was a roadmap that I, and many others, could follow—a guide to navigating the complex world of global technology leadership.

Leaders like Larry Quinlan play an integral role inspiring a generation of leaders. As a beneficiary of Larry's legacy, I recognize the responsibility to continue this cycle of inspiration and mentorship. Sharing our stories and experiences is not just recounting our journeys; it is a path for those who will follow, reaffirming that their dreams too, are within reach.

Sophia S. Ferguson currently serves as a Technology Director at Deloitte, bringing 20+ years of advanced expertise in full lifecycle software development, operations, and technology leadership. Notably acclaimed for her contributions with the "Big Four," she spearheads global transformation, formulates IT strategy, and leads global teams in delivering cross-domain technology solutions.

Inside the C-Suite: Insights and Lessons for Success

by Mario Guerendo

Learning to stand out in a crowd: My dinner with Joe Biden. Back in 2016, when he was Barack Obama's Vice President, I was invited to a holiday dinner at Joe Biden's house. In keeping with Vice-Presidential tradition, Biden was hosting a holiday party at his own residence. It was Obama's last year in office and the outcome of the election had already been decided: the Democrats had lost the re-election to Donald Trump.

The dinner party was a small gathering, no more than 35 to 40 people, and as part of the tradition, a big Christmas tree in the living room was set up for photos with the Vice President and Second Lady. Like many others, I got in line for a photo and noticed that there was a protocol person standing beside Biden. As the line moved forward, he would lean over to the VP and whisper the name of the next person coming his way; this created an intimate feeling of familiarity because when you reached the VP you didn't have to introduce yourself; Joe Biden already knew who you were.

Biden greeted me by name and as he shook my hand, asked me about a recent trip I had taken to DC. Shifting into position in front of the Christmas tree, I was standing to his right and as you can imagine, I had invested in a *really* nice suit for this event. As we waited for the photographer to move into position, Mr. Biden touched the sleeve of my suit jacket and remarked, "Hey, you should introduce me to your tailor."

I looked at him and jokingly responded, "You couldn't afford it!"

He threw his head back and laughed before saying, "You are probably right."

It was at that moment, as we were both laughing, that the photographer snapped the picture—perfect timing. I still have that photo, a memory that I will always hold as a reminder to be authentic. Make the joke. Regardless of who is in the room, be yourself.

Later that night, as I was mingling, Vice President Biden called me over and introduced me to other senior dignitaries as "the guy who said I can't afford his suit." That became a running joke and, on several occasions, provided a segue into more serious conversations about the election, politics, and other topics. I believe that seeing the VP call me by name and inviting me over into his conversations gave other guests the impression that he and I had known each other for a long time. Perhaps some folks even assumed we were tight. After that moment, for the remainder of the party, it seemed as if everyone wanted to know me—surely because it appeared that Biden and I were friends, though I also attribute it to my own ease and authenticity.

I'm comfortable navigating environments and situations like this partly because I have always been naturally confident, and partly because it's a skill I learned early and have intentionally honed over time. In family settings, I'm typically the leader, and I have learned to take on that role in business settings as well. Moving with confidence and establishing personal relationships with people is the cornerstone of progression for me—my brand. Establishing your brand as an executive is everything. It is what determines who comes to talk to you—but also what they come to talk to you about.

Stop being impressed by titles. People are people.

As I think about my journey to the C-Suite, I realize that it was my late father who instilled in me the ability to navigate in every environment with a feeling of competence and capability, so that when I enter a room (at any level) I feel like I belong there. I was raised by my single father because my mother passed away when I was quite young; my dad was a diplomat who later became an executive. He was the person who I watched all the time, even when I didn't realize it, and I picked up lessons he didn't know he was teaching. At some point in my career, I started thinking about those lessons. About how my father spoke to everyone (be they cooks or presidents) with the same level of respect. He was never impressed by anyone, no matter who they were.

I met the Pope in the 1980's, John Paul II, and that experience highlights a key lesson my father imparted. As we were entering a lounge to meet the Pope, Dad turned and said, "This is privilege. And there is nothing wrong with privilege, as long as you *know* it's a privilege." With that thought echoing in our heads, we walked in and met the Pope. I watched my dad carry on a conversation with John Paul II as he would with any normal person.

Similarly, I remember Dad engaging with the heads of state just like that—casually. The president of a country would call him with questions, and Dad was always himself, even in those situations. I can still remember him during one phone call: "Mr. President, I respectfully disagree." I think I started emulating that behavior as I became an adult, bringing those lessons along as I navigated professionally. The corporate world didn't teach me these things, I learned them from my lived experiences.

Long before the C-Suite title, people must see you as a leader.
When I think about my climb to the C-Suite, it seems to me that when I would meet various senior leaders or go into speaking engagements, even early on in my career, other people always assumed I had a bigger role than what I had. I think that's connected to the way I carry myself. When I went places with my boss, people frequently assumed I was the more senior leader. For a long time, I thought they were mistaking us (both Black men) for one another. As this continued, I realized it was the way I carried myself that led them to believe I was the boss. At some point, I accepted that I had the presence of a leader, and it was that trait that others recognized.

Lessons for aspiring C-Suiters.
Having read previous chapters in this book, it will be no surprise that I think having a mentor and a sponsor (and remembering those are two distinct roles) was critical to my ascent to the C-Suite. In my experience, I found it's better to have two separate people in these roles. The mentor is the person who often sees themselves in you; they recognize potential in you, and in many instances, they work with you to unlock that potential. The sponsor is the person who

opens doors; they know you well enough to mention your name in rooms where you are not present.

An early experience with a mentor highlights their importance. In college I studied electrical engineering, and started my career in IT doing infrastructure work, focused on switches, firewalls, and all things networking. Luckily, I had a mentor who pulled me aside and candidly shared that he thought I was putting a lot of emphasis on my technical development.

He said, "If you do all those technical trainings, you will be the guy who really knows what he is doing…but you will always be the guy who knows what he is doing. You will never be any more than that. You will be the guy who will get it done and people will trust that you will get it done right, but you will not grow beyond that. If that is where you see your career, that's fine, do that. But if you think you will want to play a bigger role, then you need to consider developing other aspects of your competency profile. You need more operational skills and soft skills."

He left me with that nugget and upon reflection I realized he was right. He worked with me and the first thing he had me do was to take a technical role in the supply chain group. Up until that point I had just been doing corporate IT and while it was a lateral move, it taught me business and allowed me to continue moving up. I slowly got away from the technical side and continued to take roles that were more operationally aligned.

Wherever you go, take all of yourself.

In my career, I have been fortunate enough to connect with leaders who opened doors. Part of what facilitated those connections is that I always aim to show up authentically. You never know which part of your story is going to resonate with people, so bring all of yourself! There are multiple points where I have opened up that have benefitted me. My international background has gotten me into meetings where I had no business having a seat at the table. I was invited into roles and opportunities simply because people knew about my background and thought I could provide value.

As an example, there was a senior direction meeting in Paris (at the time I was a technical product manager and had no business in that meeting) and the only reason I was asked to join was because the decision makers knew I had lived in France and that I spoke French.

In other situations, it could be me having lived in France; it could be that I was from Africa; it could be me having gone to Perdue; it could be me having a son who plays football; it could be me having a three-year-old daughter. We are each a collection of experiences, and you just never know what is going to resonate with people. So, bring all of you and be willing to share.

People may see things in you that you don't see in yourself. And they are right!

In my ascent to the C-Suite, it turns out I did not have to seek out a sponsor. My sponsor came to me. He said, "I want to help you," and told me that I belonged in the C-Suite. He then continued to pour into me until I believed it (and achieved it) myself. The lesson here is to surround yourself with people who see the potential in you, and most importantly, listen to them.

I wish you success and resilience as you climb. The way up is steep, and we hope this book provides insights and footholds along the journey.

Mario Guerendo is a dynamic leader with a track record of integrating disruptive technology and business strategy, known for building client-centric teams that drive digital transformation and growth. He currently holds the role of Chief Information Officer – Global Supply Chain for Cummins where he is responsible for all aspects of digitization for supply chain and manufacturing globally.

Appendix

ITSMF | INFORMATION TECHNOLOGY SENIOR MANAGEMENT FORUM

ABOUT ITSMF
Mission, Vision, History, and Leadership

OUR MISSION
ITSMF increases the representation of Black professionals at senior levels in technology, to impact organizational innovation and growth. We do this by developing and nurturing these dynamic leaders through enrichment of the mind, body, and soul.

OUR VISION
To facilitate the creation of **barrier-free** opportunities for Black technology **influencers, innovators,** and **leaders.**

OUR HISTORY
The Information Technology Senior Management Forum (ITSMF), formed in 1996, began with a conversation between a few prominent technology executives who found that only 3 percent of information technology management roles were held by Black professionals. Studies suggested that not enough Black professionals were positioned for senior-level technology positions. This discussion sparked those executives to action as they were compelled to reach, teach, and cultivate those aspiring technology professionals.

Under the leadership of Carl Williams, the 501c (3) organization was founded in Chicago, Illinois and began the work of recruiting those Black professionals who ranked among the Who's Who in technology, and those companies and individuals looking to support the mission of the organization. ITSMF, celebrating its 20th year, remains the only national organization dedicated exclusively to cultivating executive talent among Black technology professionals.

Our Leadership
Board Of Directors

Ibrahim Jackson
Chairperson

Founder & CEO
Ubiquitous Preferred Services, Inc

Dane Bamburry
Vice Chairperson

Senior Director, Enterprise Architecture
Cox Enterprises, Inc

Noreen Cohen-White
Corporate Secretary and By-Laws Chairperson

Global Technology Delivery Managing Director
Deloitte

Rona Bunn
Finance Chairperson

CIO
National Assn of Corporate Directors

Marilyn Crouther
Board Member

CEO and Principal
Crouther Consulting, LLC

Steve Edmonson
Founder & Board Member

(ret) Vice President & CIO
Cardinal Health/State and Local Government

Ehrika Gladden
Board Member at Large

President - Enterprise
Sorenson Communications

D. Lena Hickman-Miott
Board Member

Chief Information Officer and Chief Operating Officer
Miottech, LLC

Art Hopkins
Board Member

Global Head of Technology Officers Practice
Russell Reynolds Associates

Tiffany Jenkins
Board Member

Director
Dell Global Management Transition
Services Management

Fern Johnson
Board Member

(ret) CTO Vice President
PepsiCo, Inc.

Rod McDaniel
Board Member

CEO
S3 Recycling Solutions

Don Neault
Board Member

CEO, Certified Leadership and
Executive Coach, and Corporate
Advisor

Justin Reilly
Board Member at Large

CEO
Wavelo

D, K. Rogers
Board Member

Senior Manager, GBS, IT/OT Leader
and F&HC Innovation Leader
Cox Enterprises, Inc

Theresa Wilson
Board Member at Large

(ret) Executive Vice President
Wells Fargo

Michael Buckham-White
Immediate Past Chairperson

Managing Director,
Information Technology
Deloitte

Johanna R. Jones
President & CEO

Carl Williams
Chair Emeritus
(deceased)

Executive Team

Johanna R. Jones
President & CEO

Dr. Shakeer Abdullah
Co-Facilitator, Management Academy

Sherrell Bailey
Director, Events

Cheryl Bridges
Director, EMERGE Academy

Cathy Butler
Executive Assistant to the President & CEO

Cedric Cobb
Director, Development

Valda Cowan
Coordinator, College Mentoring Program

Nigel Cudjoe
Manager, GIPD Administrative Program

Dawn Jordan Jones
Manager, Brand and Communications

Dr. Rubina Malik
Director, Global Leadership Immersive Experience

Miriamm McKinney
Fractional CFO

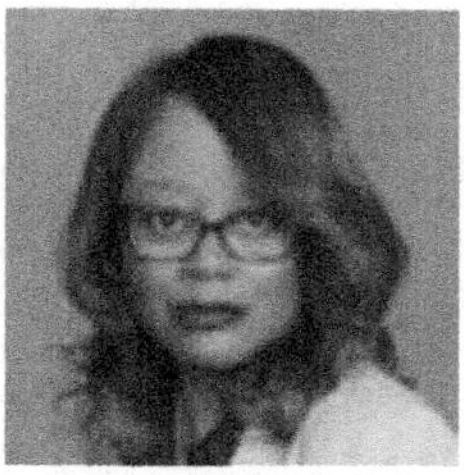

Patty Parker
Membership Contractor

Elaine Norman
Vice President of Member Services and External Affairs

Subrena Robinson
Director, EMERGE Forum and EMERGE Program Partner

Dr. James Rodgers
Director, Executive Academy

Robert D. Scott
Vice President and Dean of the Global Institute for Professional Development (GIPD)

Denise Smith
Director, IGNITE Academy

Dr. Reginald H. Turner
Director, Management Academy

Xenon Walcott
Vice President of Operations

Michael Whitley
Vice President of Development and Advancement

Made in the USA
Monee, IL
04 November 2026

67383021R00108